I0441396

Every Mind Matters:

Education - A National Priority For An American Renaissance

Nancy Driscoll

Education Priority 2

Every Mind Matters: Education – A National Priority For
An American Renaissance

Copyright © 2010

...We must not believe the many who say that only free people ought to be educated, but we should rather believe the philosophers who say that only the educated are free...

Epictetus

Preface to the 2nd Edition

Since I first provided an inside view of the crisis of public education in America and recommended a system of educational options to meet the crisis, "change" has been the mantra on the left, right, and in the middle. My position from inside the system says the "right" and the "left" are both *wrong* about what needs to be done. As "Waiting For Superman" grabs our attention, people like Oprah, Bill Gates, Jay Mathews, Michelle Rhee, Arne Duncan, and others provide a critical wake up call to Americans about the education of our children. We all want change but what we need is not a one size fits all "school system" but a "system of schools": public, private, charter, home, and virtual to meet the challenge of educating 50 million plus American children. Nothing less will do.

Having worked in public education for over 20 years I know the dedication of – and the disappointment in and disdain for – public school educators. Having taught in a private school I know the joy of a small, cohesive, classroom and the stress of raising funds to keep going. As a homeschooling mom I've experienced moments of wonder at the light of learning and seen other moms struggle along with their children. I arranged e-learning electives for a technology high school and watched in frustration as they were dropped for lack of "seat time". As a baby boomer I watch as the generation that thought we could change the world loses their jobs, their homes, and their hope.

Shall we all remain "comfortably numb" as Pink Floyd suggests? No, we have to take action. That's what Americans do. We're seeing the beginning of action today as people return to the voting booth, the town square, and the National Mall. People are writing, blogging, making videos and movies, forming advocacy groups, and getting involved. It feels like America is trying to come back, trying to regain our strength and our confidence. That's what it will take for an American Renaissance and a renaissance is what we so desperately need.

How do we keep this budding renaissance alive? How do we renew our innovative spirit, our leadership, equality of opportunity and our culture? Through the education of our children. How can an already depressed superpower regain its mojo if its children can't read. How can we compete globally if our kids can't use a globe? How can we work for the common good if we feel we don't have anything in common? Who will return jobs from India to our 9th grade dropouts? We simply must make education the national priority it deserves to be because now as never before...

Every Mind Matters

I so appreciate the support of my family as I undertook this project. I also want to thank the teachers and staff of the public schools I serve as school psychologist. I am continuously amazed at their hard work and dedication to "other people's children". Thanks also go to: all the children I've worked with over the years for adding their stories to the larger story of education in America; the homeschooling families I've come to know; the many writers, pundits, bloggers, columnists, conservatives, tea partiers, liberals, and libertarians who have shaped my views; and the many friends and colleagues who offered their insights, editorial comments, and support to this effort.

Contents

Introduction

It happened again today – another statistic in baggy jeans walks into my office at the local high school. Today it's an African American student who is failing every subject in school – tomorrow the face could be white or Hispanic but the outcome will be the same: another American child without the skills to function. It's hard to say "yes we can" when you know so many of them can't.

Dante has been recommended for a reevaluation of his intelligence and achievement levels. He's actually in school today which is rare since he is usually absent or suspended for being absent. A polite and engaging young man in "special education", we talk about school and life as testing begins. He says he usually hangs out at his brother's place then doesn't have a way to get to school the next morning. He agrees he has "anger issues" but he's "working on it". Mom would like him to do better in school but dad isn't around and step dad is a problem so he and his brother look out for each other. Yeah, I can see that.

Here we go again I think to myself as Dante quickly assembles blocks to match a pattern, analyzes concepts, and shows excellent general comprehension skills and insights. He is clearly enjoying the tasks, knowing he is doing well, and scoring easily within the average range on these critical aspects of intelligence.

Things change as we move to achievement testing. "I can't really read that well," Dante comments as he misses words at a fourth grade level. "I'm not that good at spelling," he says as we stop at third grade. "I know how to do these but I forget right now," he explains as we hit long division and fractions without a calculator. "How did I do?" he asks as we finish, "probably like seventh grade, right?" "Well no," I say, answering honestly, "more like fourth." "Damn!" says Dante, "that's pitiful."

Yes, that's indeed pitiful. It's pitiful that many American teenagers are just like Dante; intelligent, years spent in school, some even with time in special education, but still lacking in basic skills. While Dante says he wants a job working with computers, what computer jobs are available for an applicant with fourth grade skills? I see references to gang activity in his file as I look at this handsome, intelligent young man. He has the ability to go to college but the grades to be a dropout. Which will it be I wonder as I write his report with recommendations for mentoring, career exploration, and reading instruction – knowing there are no mentors, time for career exploration, or classes in high school on how to decode fourth grade words.

As we stand to leave, Dante reaches out to shake my hand and thanks me for listening and trying to help. For a moment age, gender, and race don't matter. Dante felt heard and I tried to make him believe the American dream could be his if he just worked harder. But can it? Who will teach him to read? Who will insist that he go to school and control his temper? What does life hold for an intelligent young man who has learned so little?

Given his attendance problems, I encourage Dante to commit to coming to school for the rest of the year. "Why should I?" he asks, "I'm already failing everything. I'll just start over next year." I struggle to counter his logic, knowing that three years as a ninth grader loses its appeal quickly. The bell rings as I watch him leave. "Have a good day!", he says. "You too, Dante, you too."

So surely Dante is the exception, right? As a school psychologist at a large public high school I see only the problem students. Most kids are doing just fine, right? Wrong. Take a look at the following statistics:

- Nationally 30% of our 9th graders don't graduate high school in four years.
- Nearly 2000 high school "Dropout Factories" exist around the country with graduation rates below 60%
- 70% of the nation's 8th graders don't read on grade level.
- In some states and cities, half of all students from disadvantaged groups never graduate high school at all. At all!
- America's top math students have ranked 23rd out of 29 countries on international assessments.
- The average poor, urban student enters high school four years below grade level.

No, Dante is not the exception, he's got plenty of company.

It was all the Dantes, Tonys, Jennas, Marias, Brians, and JTs – all the many hundreds and hundreds of young people I've worked with over my 28-year career as a school psychologist that prompted me to share their story. School psychologists are trained to identify and help educate children with learning and adjustment problems. We work with students, parents, teachers, school districts, and agencies to improve educational outcomes for children in schools across America. We are in a unique position to see education from multiple perspectives: the school's, the parent's, the child's – and America's.

It is this perspective and experience that made me wonder: If America realized that our educational system is not up to world class standards would they be moved to action? If they understood the personal and economic loss to America for every Dante in every school district across the country would they want to do something about it? As fears of a double dip recession – or worse - grip the country do we not see what a revitalized education system could contribute to solving the current crisis and laying the foundation for a resurgent America? Is there anything that could be said that would get parents re-engaged with their children? If the current administration, every candidate for office, indeed every American saw education as a national priority at such a critical point in our history, would the American Spirit catch fire again to demand true change – and demand it now?

Americans are restless and angry. We feel out of synch, off the mark, racing through the fog into an uncertain future. We don't like this frightening feeling of insecurity. We aren't use to it and we want change. We voted for it. We were expecting it. We don't feel we've gotten it and aren't sure who can deliver it. We've watched our retirement funds dwindle while our college grad children try for jobs at Walmart. We are tired of fighting wars but not feeling like we're winning. We are tired of offering a resume full of skills and getting no job. Americans are losing their homes, losing their jobs, and losing hope.

Will we ever feel secure again? Will we be proud again, confident in the future – our own and America's? The answer is "yes" but only if we reinvigorate the American can-do spirit in time to take charge of our own destiny as we have so many times before in our nation's history.

I recently met with a mother to share her child's test results and plan a special educational program. As we waited for the meeting to start we talked about her daughter's education –

and her own. "America has changed," she began. "My daughter wants the things other kids have but I can't afford them since I lost my job. You know, I graduated high school and I've worked all my life. I look at a job opening in the paper and think, yeah, I can do that – I've got the skills and the experience. I submit an application and never even get a call! I never even get to talk to a human being! It didn't use to be this way in America but things have sure changed. I just don't want her to have it this hard. I'll do whatever it takes for her to get a better education."

Yes, things have changed in America but we can still do something about it. In fact "we, the people" may be the only ones who can. For all the election promises, the task forces, study groups, reports, and blueprints, it is "we the people" who have to make this happen. The change we seek to renew America requires the effort and attention of every one of us. We can renew the American spirit and re-adopt its "can-do" attitude, starting in our own homes and spreading on down Main Street – the only street that really matters. Maybe even Wall Street and Pennsylvania Avenue will eventually get the message.

We must refocus on four core principles to revitalize the American Spirit. These principles need our immediate attention if we are to survive – much less reassert ourselves - as a proud nation on the 21st century world stage. Education is the key to rebuilding these four critical principles to create nothing less than a...**New American Renaissance**.

What are these principles?

Equality of opportunity

Support for freedom

Leadership in innovation

Richness in culture.

These core principles are what built the American engine of innovation and individual freedom to begin with and what will rebuild it today. Opportunity, freedom, innovation, and culture are impossible dreams without education - and education is woefully lacking in America.

So whose fault is that? Parents? Students? Teachers? The government? Yes. Yes. Yes. And Yes. It's everyone's fault. Parents hold the key to their child's education yet in large part we have abdicated our role as parents to an "electronic village". Students connected to this electronic community are totally disconnected from their own, even from their own family. Children no longer play and dream they video game. Teens live in a world of texting, sexting, and announcing their daily "status" on Facebook. Teachers put up with a system that rewards longevity rather than talent and innovation. They follow the district's policies instead of their own training and common sense. Schools create dropouts from "pushouts" since "he'll never graduate anyway". And what about the government? Don't get me started!

The Right and the Left Are Both – Wrong!

Ask those on the political right what to do about education and you get accountability, school choice, and local control. Ask those on the left and you get every toddler in preschool and spending more to "race to the top". The right says education is "up to the states". The left says it is up "to the state". Each trots out a few sound-bite-sized election initiatives and mentions opportunity and innovation as important to restore American competitiveness. But how does that happen? What changes the sound bite election promise into a proposal with an action plan and accountability for results?

Us. You and me. We must demand it. Soon our children will read about "the way things used to be" in America but they won't be living it. They'll be downsizing, cutting back, making do, and moving home. We can recharge America's batteries – but not without a world class educational system. We can't do it when intelligent children can't read. We can't do it when our teens can't find Afghanistan on a map or name the world's continents. We can't do it when a child with credit for algebra can't figure out -5 x 8 without a calculator. We can't do it when half a first grade classroom does not speak English.

Every mind matters in America – or it should. We can't accept a system which fails millions of our children, making them America's "lost generation". By looking back at America's educational accomplishments and facing today's tremendous challenges we can see what needs to be done in your house, the school house, the state house, and the White House. Let's get busy.

Education Priority 16

...Education is the transmission of civilization...

Ariel and Will Durant

Every Mind Matters:
Education Is Critical To America's Renaissance

1

The American Spirit

America holds a unique place in history. Its experiment in self-government and individual freedom has made it the envy of the world for the last 200 years. Americans have always been known as "rugged individualists" with early American author James Fennimore Cooper creating the genre of the "American Hero", capturing the imagination of a new nation and the fascination of the old.

As America grew into the 20th century the world was changing too. The rule of kings and queens was giving way to the rule of law and democracy. As noted in Peter Jennings' *The Century for Young People,*[1] at the turn of the 20th Century American inventors began to thrill the world with their machines, electric lights came on with the flick of a switch, and the first American "skyscrapers" rose aloft. Jennings writes of a young Russian immigrant approaching New York Harbor's Statue of Liberty for the first time to see everything "big and new" and "full of hope for a beautiful life in a new nation". Like nearly thirteen million others who came to America at the turn of the 20th century, this young man came to live the American Dream.

...they came full of hope for a beautiful life...

The American Century

"The American Century", as the 20th century came to be known, was a time when "the American spirit" meant innovation, freedom, and limitless possibilities as people from all over the world came together to build a nation. Picture these scenes unfolding across a newly re-united nation, healing from the crisis of civil war and finding its place in the world as it headed into the 20th century:

- An iconic American cowboy sits on horseback sipping hot coffee while he watches over a herd of cattle in the frost of an early spring morning in the high country.

- The green legs of a John Deere tractor pick at rich black earth on an Ohio farm, crawling up and down golden fields of wheat to feed the world.

- A high school band strikes up the Star Spangled Banner in the Friday night lights of a football field as the bleachers fill with moms and dads, aunts and uncles, eager to cheer their hometown heroes.

- Grandma readies warm apple pie for the family as dad returns from his shift at the auto plant.

- Raucous convention debates finally yield a presidential candidate and national platform as a nation waits to elect the next leader of the free world.

- Steel workers in Gary, Indiana forge the building blocks of the world economy – or at least they used to.

The scenes described above are history - just some of the "good old days" every body talks about. Why? What happened to the American spirit? Why don't we feel as good about ourselves and about our place in the world as we used to? Why does it seem that nobody likes us anymore? What does "American" even mean?

We were "united", now we're "red", or "blue" or "tea party".

Instead of the American family tradition we now have African Americans, Hispanic Americans, and Asian Americans. Instead of "united states" we are merely blue states and red states on a CNN map. We no longer feed the world, the world feeds us – and we're afraid to know what's in it. "Made in Japan" is a source of pride as Toyota became the largest car maker in the world and we the people bailed out Chrysler and Chevy. The steel plants are closed, grandma lives in a retirement community in Florida, and X-rated movies entertain us instead of westerns.

And what of the educational system that used to be the envy of the world?

America led a world-wide technological revolution but now our students' achievement has fallen well below our competitors around the world. In some of our cities half of the students never graduate – yes, they never graduate high school! Parents must be reminded by research studies that it's good to have a family dinner – at least once in a while!

Even the Europeans want to know what has happened to the upstart young country across the pond. The Nov. 27, 2007 edition of the German news magazine, *Spiegel*, wonders about "The Depressed Superpower"[2]. They note that frustration has taken hold in the land of optimism, making Americans seem more like – well – Germans!

...what's gone wrong in the land of optimism?...

They write that Americans are beginning to doubt their greatness. And why not, Spiegel asks, with wars in Iraq and Afghanistan, Washington DC school students without books and afraid to walk the halls, no more "made in American" labels, and an aging baby boomer population that has lost its determination to change the world. Perhaps the great "rock philosopher", Pink Floyd, said it best: "We've become comfortably numb". Perhaps we don't even know what we don't know. We don't even realize what we've lost. We don't know that math and science might as well be foreign languages to many American school children.

Spiegel comments that optimism used to be part of America's genetic makeup. Today demographers have begun to paint a picture of America as a melancholy, dissatisfied, and bitter nation. Are they right? Reportedly 60% of Americans now believe that the next generation will be worse off than their own and that our country is headed in the wrong direction. A majority of us have little confidence in our government's ability to solve the nation's problems, economic or otherwise. Historically, most of us don't even vote. Hope was ignited as voters re-emerged in force for the historic election of Barack Obama. Today those voters may stay home while others are emboldened to demand new leaders. Is it "Yes, we can", "Well, maybe we can't", or "Can anybody"?

According to London Times writer Matthew Parris in "Rusty superpower in need of careful driver"[3], it will be "No, we can't". He writes that President Obama will manage *not America's*

renaissance but its national decline – a hard job he believes
since Americans don't believe they can be beaten at anything.

Have we already been beaten? Have we already given up the
fight? Will the younger generation's enthusiasm for change
and the excitement of the election of the nation's first African
American president help us boomers remember what it was
like to march, protest, and demand the changes we achieved in
the 60s. Will we find common cause in a tea party revolution?
Are we willing to organize to save the next generation?

**Children spend hours and days and years of their lives
"plugged in" to the electronic village.**

America has always been at the leading edge in developing
new technology. American innovations and technology
changed the world for the better, right? That's debatable.
Brave New World, 1984, and *Fahrenheit 451* have slipped up
on us with the click of the remote. No need to argue about
whether it takes a family or a village to raise a child. We're
letting our children grow up in an *electronic* village. Our
children spend *hours and days and years* of their lives in front
of the TV, video screen, cell phone, and computer. We are
watched from every street corner and department store. Our
computer knows our name and our shopping preferences. Our
history gets a more politically correct slant. Books and
newspapers sit gathering dust on bookstore shelves as
everything becomes digitized. As my own daughter sat reading
The Lord of the Rings her friend commented, "You know that's
out in DVD, right?" We communicate with emoticons and
tweets and wonder why our children can't write a complete and
coherent sentence. On a recent trip to Disney World, children
and teens were "experiencing the magic" with headphones on!
Why even go?

Americans can handle anything but a loss of hope.

From the German perspective, Americans can handle anything but a loss of hope. *Spiegel* notes American journalist and author William Safire says that if government institutions are the skeleton of the body politic, the American Dream is the soul. If the dream dies will the body not follow? *Spiegel* observes that Americans want change yet we are afraid of it. We want the government to intervene when it suits us but not when it doesn't. In the last election politicians took up the mantra of change and voters responded. Now we see bumper stickers suggesting "You keep the change!" The Tea Party movement is also changing the political scene, offering a voice to those dissatisfied with the changes they see.

In his book, *What Went Wrong With America and How To Fix It*,[4] talk show host Darrell Ankarlo heard a lot of what's wrong with America from his listeners and their ideas of what needs to change (italics mine):

...we've lost trust in our leaders...

...*the failure to teach our children morals, honor, integrity and personal responsibility*...

...we're throwing away the republic on which we stand...

...we lost the notion of right and wrong...

...*the decline in parental authority*...

...*we've lowered our standards*...

...immorality is made to look like the norm...

...fast-paced lifestyles...keep us from communicating with friends, family, and neighbors...

...we've become convinced we can't take care of our children, vote, or anything else without government intervention...

He also heard about what needs to change to fix what's wrong – education, family, and activism:

...people must be more educated, more informed, and more willing to act...

...take the blame and make a stand...*make your voice heard and encourage others to do the same...*

...turn off "the box" and eat, talk, and dream together as a family...

..become an activist...stand together and make ourselves heard...

Americans recognize that change starts with the actions of each of us.

Yes, America faces challenges, but are these challenges so much more daunting than those of the past? Do they surpass taming a nation, surviving our own civil war, emerging from the depression to become a super power, or marching all the way to Washington DC for civil rights? As its biggest age cohort, the Baby Boom generation, ages is America itself at its mid-life crisis? Are we becoming less courageous, less confident, a little unsteady on our feet, less willing to take risks and too tired to care? Could there be a march on Washington today for a cause like our children's future or would Americans prefer to watch one on TV instead? The annual Right To Life march brings nearly a quarter million Americans to its capital every year yet rarely gets a mention on national TV. Is anyone out there? Anyone interested? Willing to stand up and be heard?

Education For An American Renaissance and a New American Century

Ronald Reagan once evoked the image of "morning again in America" to try to wake us from our doldrums after Viet Nam and the oil challenges of the 70s. We re-inflated our egos in the go-go 80s but the bubble has burst again and it feels like the sun is setting on the American Century as it did on the British Empire of the 1800s and the Roman Empire centuries before that. Is it too late for America? Must President Obama merely manage our inevitable decline? Perhaps morning only dawns in India and China as the Asian Century begins.

What would life be like for your children in an "Asian Century"?

The British too wonder where the next century is dawning. In "Why China is the Real Master of the Universe", Anthony Browne ponders the global shift of power from West to East[5.] He notes that as the baton was passed from England to America in the 20th century, so now the baton is being passed from the West to the East for the 21st century as China and India take the leading role on the world's stage. The Chinese are now being dubbed the "New Colonists" as they seek influence, raw materials, farm land, and economic clout around the world. Whereas the British and American centuries shared a language, rule of law, and cultural heritage, the Asian Century offers a far less predictable future for our children and grandchildren.

As a British economist recently noted in a CNBC interview: "In 1832, China and India were the world's two largest economies and by 2032, they will regain that status." The decline of empires happens much faster than we think.

2

Goodbye America, Hello... Asia?

In a fascinating and very readable book, *The Elephant And The Dragon: The Rise of India and China and What It Means for All of Us*[6], Forbes Foreign Correspondent Robyn Meredith explores the rise of India and China on the world stage. She looks at globalism, offshoring, and free trade policies that are having a world- and life-changing impact on America and every American. While that does not come as news to anyone who has lost a job to cheaper labor overseas, her analysis and the policy suggestions she makes remind us that America must quickly regroup if our children are to experience the American dream. You have to take charge of your child's future and demand that education become a national priority. It's not just that the world is changing. Your own little corner of it is changing too.

The world's poor want your child's job!

Through globalization and free trade the world's poor have benefited from new jobs and an improved standard of living at the expense of middle America. Indeed millions of people in China, India, and other nations have been delivered from dire poverty as jobs and commerce flow more freely throughout the world. The displacement of workers is certainly nothing new, the author points out. The industrial revolution, assembly line manufacturing, the IT revolution, all led to the elimination of people's jobs. Job gains for some mean job losses for others.

The New Silk Road

Cheap manufacturing costs lured business to China. Today 75% of all new toys are made in China – prompting concerns about toy safety. One out of every three pairs of shoes in the world comes from China. China has also invested heavily in its infrastructure and in doing so is suffering significant environmental damage while foraging the world to meet its vast appetite for natural resources and buying up farmland on other continents to feed its people. Hosting the 2008 Olympics with a spectacular opening ceremony showcased modern China's reemergence on the world's stage – perhaps its own renaissance. And not only can China produce, it can buy. With new jobs comes purchasing power as an eventual 100 million Chinese attain middle class status. American companies cut costs by moving manufacturing to China and reach an entirely new customer base at the same time. Still the ever-expanding march of globalism and the impact of the global economic crisis continue.

...100 million Chinese will attain middle class status...

As China's workers demand higher pay and better working conditions, the world's manufacturers set up shop in a next-tier country such as Viet Nam or Cambodia where a new mass of impoverished workers awaits the promise of the global marketplace. As American consumers cut back spending, Chinese workers lose jobs and head back to the impoverished countryside. Korean mothers take their children to America to be educated and learn English so they can return home to compete with children in China and Japan. Alarmed that its own students are falling behind South Korea and Hong Kong, the Japanese are adding 1200 pages to elementary school textbooks and "going back to basics" after a 10-year experiment with "stress-free education.[7] Science and math textbooks are getting 60% more pages, time is being added to the school day, and English instruction will begin in fifth grade. This is your child's competition. Does he or she even have a science textbook? What about a foreign language? Is he or she ready for an Asian Century?

And The New Spice Route.

India too has arrived as an economic super power as its well-educated citizens manage the world's information. They remain behind China in many regards, including manufacturing capacity and infrastructure, but they are committed to continued growth. Even poor families scrape together money to get their children into school. "The Beautiful Tree" by James Tooley recounts tiny rooms full of students in India's slums whose parents scrape together tuition to give their child a better opportunity[8]. Those children learn English, science, and math to attain middle class status in jobs at wages far below those your children will expect.

China and India were global powers centuries ago when the world sailed to them for silks and spices during the age of exploration. China is no longer a country of invention as it was when it brought the first paper and porcelain to the world. It now takes America's inventions and builds them cheaper and faster while India keeps track of all the transactions.

The other applicant for the job your child wants may be half a world away, better educated, and willing to work for half the salary. Again – how will your child compete?

Well at least your child will be training for a white collar job, right? Your son or daughter won't be competing with Chinese students to make gym shoes, right? Probably not, but they will be competing with children in India for jobs in many other fields. India has already received more than a million white collar jobs including those in call centers, programming, accounting, investment banking, data management, and information processing. Even higher numbers of service jobs have gone to India. Ask any laid off white-collar worker where his or her job has gone and the answer will likely be "India". American technology companies are investing heavily in the well-educated, English-speaking, cost-effective, Indians – to the tune of billions of dollars – in support of the world's future workforce. So who exactly is investing in your child's future?

...What are the silks and spices of the 21st Century?...

What silks and spices can America offer the world marketplace today?

Certainly the news is not all bad. The world's new middle class will be able to purchase American goods we are told. That will create new jobs for our children we're assured. That's the benefit of free trade. But what do we have to offer the world's new middle class?

What are the silks and spices of the 21st century with which Americans can compete? If we cannot compete in manufacturing goods or managing information, where can we be competitive? We are exporting American culture but is that the face of America we want to present to the world? Will the world continue to clamor as they have in the past for all things "American" from our jeans to our music or will Asian culture proudly return to predominate?

While American factory workers used to efficiently assemble every piece of every product from computers to cars, today's new workers assemble a single component and send it on to the next worker – often on another continent – for final assembly. After dreaming up the computer and creating the internet, will America's children be relegated to texting and "Facebook" while Indian workers keep the world's information moving? We invented the automobile but will our children be able to afford to own one or pay for the gas? Even companies trying to return jobs to America have a tough time. AT&T CEO Randall Stephenson noted that only 1400 of the 5000 service jobs he was trying to bring back to the U.S. from India could be filled.[9] The company couldn't find the skilled workers it needed in the U.S.!

American education is not getting the job done.

Ms. Meredith provides the thoughts of some of America's top thinkers as well as her own insights into what is needed to re-invent America for us and our children. A review of those insights provides the background for my views as an educator and parent. We must awaken the American spirit for our children by realizing that education is the key to a new American Renaissance (italics mine):

- There are more college graduates each year from China and India than from America and Europe combined – *your child has global competition.*

- A larger percentage of Americans than Chinese go to college - but *a smaller percentage graduates.*

- India has about 2.7 million college graduates each year, about twice that of the U.S. A *larger percentage of these graduates are engineers* compared to the U.S.

- *American businessmen note that American education is failing to meet the needs of its citizens.*

- Spending on basic research in America has been cut back when *innovation may be the most promising American commodity* (think silks and spices) *for the 21st century.*

- Stephen Roach, chief economist of Morgan Stanley, says that *re-inventing ourselves is the only way forward for America.*

- We must *create new types of jobs that can't be exported* or our children will be worse off than we are.

- Robert Rubin of the Hamilton Project at the Brookings Institute notes that we *have to have an education system that is "first-rate"*; we have to put in place the ability *to be the best we can be* again.

- *Americans can once again become rich in spirit, creativity, and innovation.* To do that *we must strengthen the most critical building block – K-12 education.*

- Craig Barrett, chairman of Intel, says that we *must fix American schools*, emphasize math and science, and *parents must demand that their children take rigorous courses.*

- *Feed innovation with education* and research to create the jobs of the future.

- If our government can't or won't meet these challenges, *American families must take on the responsibility for their own children.*

- Encourage children to look at "green" careers (more silks and spices), high-end services, and "person to person" jobs that can't be exported.

- Infosys CEO Nandan Nilekani feels America will meet the challenge as we have many others, but notes we have to focus on education. *"The education investment is a must, it is nonnegotiable,"* he says.

- Let the return of America's competitiveness be *"this generation's space race"* notes Ms. Meredith in conclusion.

This Generation's Space Race

As a member of the baby boom generation, I remember watching on TV as Neil Armstrong took that "one, small", first step on the moon for all mankind. Mr. Armstrong was from my hometown and everyone I knew was in front of a TV, waiting to see the moon landing. What a moment to witness! President Kennedy told us we could do it and we believed him. America had gone to the moon and a kid who grew up in a little town like mine had become an American hero. Today we are told we simply can't afford to go back. Unbelieveable!

When America set a goal – no matter how difficult – we marshalled all our resources and reached it. We won the world's wars, now we simply cease fire. We freed the enslaved, now we move on. We revolutionized science and technology, flew to the moon and set our sights on Mars. We can do it again but we need to believe in ourselves, believe in the American spirit. Remember: the first thing to do to reach your dreams is to wake up!

Education Priority 32

As Ms. Meredith notes:

**Let the return of the American spirit be this generation's
space race.**

*Education is the most powerful weapon which you can use to
change the world...*

Nelson Mandela

3

A New American Renaissance of Education and Innovation

Goooood Morrrnnning, Ammmerica!

The Alarm Has Sounded. It's Time To Wake Up!

Surely the sun hasn't yet set on the American dream. We need not dominate the world as its policeman but we need to lead and share the ideals we believe in. We need not dictate to the world's economies but we can surely work together to improve our own. We can help lift up the poor around the world but we must attend to our own poor as well. We can still be a force for what is good and right and just. The world needs us to wake up, and so do our children.

I am proposing a re-examination of four great principles of the American Spirit that need our immediate attention if we are to be a force for good in the world and insure a happy and prosperous future for our children:

- Equality of Opportunity,
- the Championing of Freedom,
- Leadership in Innovation, and
- a Richness of Culture.

We already have the foundation on which these principles are built. It's called the Constitution and the Bill of Rights. It's a solid foundation that has stood many tests of time. Who's to undertake such a huge task? Us. You and me. The Greatest

Generation. The Baby Boomers. GenX and Y. Our children and grandchildren. All of us together can make it happen.

The greatest generation gave us a respite from war, a work ethic, and moral values that were grounded in a core belief system.

The baby boomers questioned those beliefs and values, fought unpopular wars, correctly demanded equal rights for all citizens, then used some of those rights in a selfish attempt to indulge themselves as they had been indulged, becoming a silent and apathetic majority.

The GenXers may be cycling back to the middle with interests in more than their own needs and wants as they become faced with the challenges of starting families and building careers in a changing and frightening world economy.

Today's graduates stare at student loan bills, move home with mom and dad, or consider grad school when their degrees don't bring jobs and the change they worked for hasn't materialized.

Each of these generations shares the stage today. Our history is America's history. We are making tomorrow's memories for our children. What we do today will shape America for the generations that follow.

My generation dreamed about changing the world – making things right – when we got the power "to the people". I still remember walking across the graduation stage of my small high school in Ohio, knowing we couldn't trust anyone over thirty to straighten out the crises of Viet Nam and civil rights so we'd have to do it ourselves. I remember campus riots and Kent State. I remember flower power and peace symbols.

We had a "we can change the world" attitude and despite the obstacles we did it.

So who do we trust today to get things done for America - Wall Street or Main Street? Can we trust the State House or the White House or should we look around our own houses for answers? The "change the world" mentality of my generation is still part of our psyche, it just needs renewal. We clearly have the *numbers*. Now if we just had the *will* we could be a silent majority no more and join our grandparents and our children to rebuild the American dream.

Education, The Common Thread.

What is the common thread that is needed to pull this all together? *Education*.

Education, the great equalizer.
Education, the foundation of responsible freedom.
Education, the seed of innovation and job creation.
Education, the nurturer of culture.

Education simply must be Priority 1 in America.

Renewing the Principles

Why are these principles important and why is education the key to rebuilding them? Opportunity, Freedom, Leadership, Culture. They are important because they are what inspired the first American Century. They are what brought people from all over the world to our shores – and why the next generation of immigrants is still trying to get in. They are why the very name "America" used to inspire such hope, respect, and pride and made us the envy of those who had been denied.

Consider why each is critical for a renaissance of spirit in America through education – why they make us so uniquely American.

Equality in opportunity

There has been an historic belief in America that with education, hard work, and perseverance anyone could succeed. Give me "40 acres and a mule" and I'll become the world's breadbasket. "Pull yourself up by your bootstraps," those who are down and out are told. "Start at the bottom and work your way up."

...Are we still the land of opportunity?...

The millions of people who have come to America over the past decades have come for this very reason. They saw America as a place where through education and hard work the "American Dream" could be theirs. Perhaps nowhere on earth did such equality of opportunity exist. Our Statue of Liberty still "beckons those yearning to be free" as it did the millions who came across the oceans at the turn of the 20th century, yet today's immigrant faces a very different America. As opportunities become more limited the land of opportunity wants to tighten – maybe even build walls around - its borders.

Clearly, opportunities in America have not always been equal and, some would argue, still aren't. Perfect equality of opportunity is still an unrealized goal but through the will of the people doors have been opened through education to women, minorities, and the disabled that have leveled the playing field. America has been an example to the world in its belief that any barrier can – and should – be overcome.

Champions of freedom

America becomes "American"

The 20th century - "The American Century" - saw America grow out of its bitter civil war into a champion of freedom around the world. We became less "yankee" or "rebel" or "frontiersman" and more "American". Touting our own experiment in education for all, democracy, and freedom of opportunity, we were sensitive to the threats to freedom faced by other countries around the world. This was especially true of Europe, from which so many Americans could trace their own heritage.

Ignoring George Washington's plea to refrain from European entanglements, we twice entered the wars of Europe – now known as "world wars" – to ensure the freedom of others. As the World War I song went, "the yanks are coming, the yanks are coming, and we won't come back 'till it's over, over there".

The "greatest generation" of yanks came to the rescue again in the 1940s after our own country was attacked at Pearl Harbor. We were hailed as liberators when town after town, country after country, was pulled from the influence of Nazi ideology. What might Europe, and indeed the world, look like today had America not stood with its allies in support of freedom?

... We won't come back 'till it's over – over there, but do they still want us – over there?...

Of course our defense of freedom can and has been interpreted as meddling in the affairs of other countries and as nation building, and our successes have been tempered with new threats of terrorism. The baby boom generation fought in unpopular wars in the far reaches of the world while the current generation's soldiers find themselves in the deserts of the Middle East. We fought a cold war with communism that saw the countries of Eastern Europe fall behind an "iron curtain" until the right mix of leadership here and abroad brought that curtain down. Today we fight unseen, even "homegrown", enemies intent on punishing us for perceived wrongs.

We use our economic might and rhetoric to try to impact dictatorships in Central and South America, or at least we used to. We decry human rights violations around the world but that outcry is tempered by our need for trade. Are we still viewed as champions of freedom or as the bully on the world's block? Only an educated America can champion freedom and speak with moral authority to improve the lives of others.

Leaders in innovation

Along with the surge in immigration to America at the turn of the 20th century came life- and world-changing technological innovations:

- The telephone enabled instant communication across a continent and eventually around the world.

- Radio and the movies began to create a shared American culture. People all across the country were listening to the same radio shows and watching the same movies. Hit songs, fashions, and dances took on the description "American".

- American aviation opened the skies for travel as the Wright Brothers took their first flight.

- The assembly line enabled Americans to mass produce products with an assurance of quality. "Made in America" became synonymous with quality while "Made In Japan" meant just the opposite.

- The advent of electricity lit up the night.

- The first "Model T" was built in 1908 and by 1920, eight million cars were rumbling down the roads of America. The American automobile was the envy of all and everybody wanted one. Buying "on credit" provided a way to get you what you wanted and get it for you now.

- From computers to medicine to agriculture to laser technology, American innovation changed the world

Where is American innovation today? Where are the great discoveries, the new technologies, the world-changing ideas? Whose children are dreaming up the future? Probably not yours.

Rich in culture

Clearly, America is a nation of immigrants. Europeans began arriving on the continent hundreds of years ago to share its beauty and wealth of natural resources with its native inhabitants. Of course those native Americans would offer a very different perspective as they were systematically driven back across the continent to make way for a new American who worked to tame the wild country.

Predominantly settled by the English, French, and Spanish, what is now America also drew other Europeans fleeing the continental wars, famines, and religious persecution of the old country. Those who came were by nature opportunists –

confident enough in their own abilities to strike out across an ocean to do their own nation building.

Others came not of their own will, but chained together on slave ships from Africa. Much of America was built on the backs of these Africans who became "African Americans" and would have to fight for their freedom and rights as human beings. Eventually freed by the force of America's Constitution, these new Americans went on to start businesses, invent, and create their way into the American story.

The "American Experiment" as it became known would prove to the world that individual freedoms were inalienable and that people of different beliefs could eventually live and govern together peacefully. Dissenting views were to be protected with specific rights divinely – yes divinely - endowed and guaranteed to each and every American.

...Americans – opportunists by nature...

While the new nation shared common beliefs about liberty and freedom, the comforting traditions of home were maintained as well. From Mardi Gras in New Orleans to St. Patrick's Day parades in Boston, each region of the country offered its own flavor of Americana. Jews, Catholics, Russians, Africans, Chinese, Irish, Polish, Italian – all found their place while building America through education and their work ethic – creating their own American dream. From the Scandinavian influence in the upper Midwest to the Spanish heritage of the southwest, America's rich mix of flavors, talents, and perspectives made the nation the envy of the world. How do we reawaken that spirit of America for our children?

4

Making Education... Priority One

Reawakening The Spirit

How do we reawaken the American Spirit? How do we come out of the lingering recession to restore some sense of confidence in the future? How do we regain leadership in innovation? How do we ensure equality of opportunity? How do we celebrate the richness of culture without losing what makes us all American? How do we promote freedom here and abroad? How do we refocus on the principles upon which the American spirit depends?

- Educate our children.
- Teach them the American story.
- Show them the path to leadership in a world facing many problems.
- Create new opportunities through innovation.
- Celebrate our shared culture.
- Convince them that education equals opportunity.
- Understand that "every mind matters".

Education. It Simply Must Be Priority One In America.

Education and Equal Opportunity

Every mind matters – to all of us. Just as the slogan of the United Negro College Fund has pointed out for many years, "a mind is a terrible thing to waste". Education has always been viewed as the ticket out of poverty, the pass to a better life, the leveler of playing fields, the one thing no one can take away from you.

As I work with parents and children in schools, it is clear that most parents care deeply about their child's education. I hear the stress in their voices as a committee of educators asks for "parent concerns" to develop an intervention plan for a student in trouble.

- "She hates to read."
- "Homework is a battle every night."
- "Do you think he'll pass his grade?"
- "I worry about him in middle school."
- "I don't want her to drop out."

Every parent seems to know that without a good education, their child's life will be filled with missed opportunities. Many parents say "I didn't do good in school myself." or "She needs an education to get a job." That parent may not have graduated, been able to go to school, or gotten the job offered to a better qualified worker on the other side of the world. They want something better for their own child and know education is the ticket.

Education and Freedom

Only an educated populace can remain free. One of the key goals of education in America has historically been to educate citizens for full participation in the process of democracy. An educated citizen can make informed decisions, resist tyranny, analyze promises and proposals of the government, and be part of solutions. He can hold politicians accountable, work and pay taxes, and contribute to the civic good.

As I talk with and evaluate students it is always sad to document what they don't know about America and how it works. "Who was president during the Civil War? – Don't know." "What are the three branches of government? – Don't know." Ask your own teen the following questions as a sort of...

"Freedom Quiz":

1. Name three guarantees in the Bill of Rights.

2. During what time period was the Civil War fought?

3. From what country did America become independent after the Revolutionary War?

4. Why do we have both a Senate and a House of Representatives in Congress?

5. Why did the Founding Fathers think a "separation of powers" was needed?

6. What groups of Americans were once not allowed to vote? What happened to allow them to vote?

7. What would you say is the main responsibility of the federal government?

8. Name four countries where America currently has troops stationed and explain why we are there.

9. What do federal taxes pay for? State taxes? Local taxes?

10. Who is the governor of your state?

So how did they (or perhaps you) do? When I ask students about their favorite subject, "history" is never the answer! "It's boring", "It's just a bunch of old people and dates", "It's so over" are some of the responses. Yet those time-worn phrases of history repeating itself and us being doomed to repeat our mistakes are still true. If today's children have no idea how and why America came to be, how can they preserve America's ideals and take us into the future?

Education for Innovation

We need to create a New American Century of leadership and innovation that is nothing short of an American Renaissance. Workers in America now compete with workers in every corner of the globe, as noted by the National Center on Education and the Economy[10]. As anyone who has called a helpdesk knows, the people on the other end of the phone, even if they call themselves Susie or Bob, are likely to be sitting in a cubicle on the other side of the world in the middle of the night.

Banking, commerce, IT, any field of endeavor can find and employ an educated workforce wherever labor costs are lowest for comparable quality. Businesses decry the lack of science and technology graduates in America as they happily hire foreign workers at lower wages. Indeed Americans are no longer assumed to be the better educated employees. "American" as an adjective used to be synonymous with "quality", "high-tech", "well-educated", "innovative", etc. Now India and China are rising to the challenge, claiming the turf that only America used to occupy.

... Made in America? I don't think so...

China and India are turning out more mathematicians, scientists, and engineers than America. In fact two other subjects that the children I work with never rank among their favorites are science and math! I often hear "I'm just not good at math" and "I'd like science better if I could *do* it". Why should math be so hard for American students? What *is* science if not *doing* things to see what happens? Sometimes I do get a high school student who responds to my "what kind of career are you thinking about" question by saying "engineering!"

The following conversation is typical:

"Great!" I say, "What do engineers do?"

"Uh, I'm not really sure," says my student.

"Well, what graduation pathway are you on, college prep?"

"I don't know."

"Hmmm, well how are your grades?"

"I think I'm passing."

"You think you are?"

"Well, I know I passed a couple classes."

"Well, let's see. You're 16 now. How about Freshman English and Algebra?"

"I think I passed English but I haven't passed Algebra yet."

"You're still taking Algebra?"

"Yeah, this is my third time..."

"Well, OK…, where would you like to go to school?"

"Duke!"

The disconnect for some children between where they are (if they know where they are) and where they want to go is frightening. What jobs does America hold for these children? What is their role in the global economy? Are the prospects for these – our own American children – really much brighter than the poor of resurging China and India?

Education and Culture

Melting pot, salad bowl, or buffet?

At the turn of the last century, America was known as a melting pot of cultures, languages, and ethnicities. The idea of a "melting pot" was that these cultural differences were stirred together to form a richer – but now "American" – stew.

More recently we hear America called a salad bowl. Each "ingredient" is distinct yet blends together to make a nutritious offering – or so we hope.

Today it seems the analogy of a buffet is more appropriate. Americans do love a buffet! We pick and choose. We put on and take off. We steer clear of what we don't like and pile on what we do. We want to retain our traditions not only in our own homes but to observe them publicly as well. We clash over where my rights end and yours begin. If it is OK for me to wear a crucifix why can't you wear a headscarf? If schools and businesses close for Christmas why don't they close for Jewish holidays or allow bells to call believers at the appointed times? I resent you not knowing English and you demand a driver's license test in your own language.

And what of the culture we export? Are Americans defined in the world by the X-rated movies, violent video games, and degrading rap music we produce? How can we speak with moral conviction for human rights and freedom? Would you listen to us? And what are we feeding our own children? If your children watch network TV they see murder, adultery, cheating, adult-themed comedy, and more every night – never mind what they can get on cable.

Murder and Mayhem for 6-year-olds?

I provide counseling groups to elementary school students. I'm shocked as they reference the most violent movies and video games as I try to help them identify their feelings and control their impulses. As we talk about what scares them it isn't a blue monster under the bed or a friendly ghost. They're scared by video games involving murder and mayhem – and they're six years old! That monster under the bed that daddy used to scare away is now an electronic devil in the latest video game. As they tell me about how a character's head got blown off or his heart torn out I wonder if it's a losing battle to teach them not to hit each other over who took their snack! How can we allow our children's minds to be filled with such violent and often sexual images? Who are the adults here? Will you please raise your hands!

Opportunity, Freedom, Innovation, Culture

What does the future hold for our children without a much better quality of education? Not The Benefits of The Four Principles Of The American Experience:

Opportunity: a good chance

Freedom: not being under another's control

Innovation: new ways of doing things

Culture: customs and arts that define a nation or civilization

We know what we want and need to accomplish. We want our children to have every good chance, not to be under anyone's control, to find new ways of solving the world's problems, and to create a culture our nation can be proud of when our descendents read about America centuries from now. Renewing the four principles requires a re-commitment to education but how is the American educational system doing?

We were a nation "at risk" in 1980. Have we fixed public education in the last thirty years of trying? Not according to a Ed In 08 report on the 25th anniversary of the "Nation At Risk Report". According to "A Stagnant Nation: Why American Students Are Still At Risk"[11], America still got an "F" on raising standards, providing more time for instruction, and tying teacher pay to student outcomes. Everyone from Wall Street to Main Street says education needs fixing so what exactly is wrong with it?

In 1998 the US ranked first in the world in the percentage of young adults with a bachelor's degree. By 2004 we had dropped to fifth. Our 8th grade math curriculum standards are two years behind those of other countries. Out of 29 countries participating in educational assessments in 2003, American students ranked 24th in math, 24th in problem solving, 18th in science, and 15th in reading! Even controlling for student ability we were well behind our peers around the world with America's top math students still ranking 23rd compared to their peers in the 29 participating countries. The most recently released 2007 report of the international math and science study shows American students improving but still lagging other students in top-scoring countries. How can America recover its economic footing when its children don't have world-class skills in such basic and critical subjects?

Education is clearly the engine for job creation in America. If we could raise students' skill levels to the *average* level in Europe, yes we are well behind the average of Europe, it is estimated we could increase our Gross Domestic Product 5% over 30 years. The resulting job creation and income potential is staggering in comparison to the costs required to attain those educational benchmarks. The Organization for Economic Cooperation and Development[12] suggests low education standards as one of the biggest threats to the US economy. They note the US has not only lost its lead in education but that the quicker progress of other countries is likely to widen the gap.

...The average poor and urban American student enters high school up to four years behind grade level in reading and math. Do you think they catch up?...

According to a Strong American Schools report[13], only one out of ten Black eighth graders reads at a proficient level and as they near graduation Black and Latino students have skills no higher than white middle school students.

Are these statistics really any worse than before? Aren't there always those who "can" and those who "can't" in schools? Isn't someone always moaning about the state of education to get more money? Isn't this just the latest crisis in education? Other countries don't even try to educate everyone do they? Let's just accept that not everyone can succeed.

Yes, but not even most students are succeeding...

- *8th grade math curriculum two years behind other countries*
- *Dropped to 5th in bachelor's degrees*
- *24th out of 29 countries in math*
- *24th in problem solving*
- *1.3 million drop out each year*
- *18th in science*
- *15th in reading*
- *our top math students are 23rd*
- *we are still behind the top countries*
- *how can our children compete?*

Maybe everyone from the kindergarten parent next door to the CEOs of global companies is just crying wolf – but do you really think so? Maybe a few million uneducated minds don't really matter that much. Or do they?

...Many public school children seem to know only two dates,

1492 and the 4th of July; and as a rule they don't know what

happened on either occasion...

Mark Twain

5

A Brief "Been There, Done That" History of American Education

A quick look back at American education reminds us there is really not much new under the sun. PTA fundraisers? Been done for decades. Homeschooling? Used to be that was the only way to learn. Lack of parent involvement? Been there, complained about that – for the last 100 years!

In colonial America, few children went to public schools as we know them today. Most children were educated at home. The privileged few attended private schools. The family retained primary responsibility for the moral and cognitive development of children.

During the 19th century, as education became more formalized, some tension began to grow between home and school regarding how best to educate the young. While rural parents maintained more control over their own small schools and teachers, in the urban centers educators became more organized, confident, and vocal about their profession.

By mid-century the parent's role in education was in decline while the role of the professional educator was becoming more defined. As noted by William Cutler in *Parents and Schools*,[14] both the family and the school were "institutions in transition" during the 19th Century.

The First PTAs: Home-School Leagues Are Formed

Even in *1896* educators observed that parents believed "the teacher is appointed to relieve them of their parental and natural task of instructing their offspring!". Sounds like a teacher in *1996* complaining about a lack of parent involvement, doesn't it? By 1900 educational reformers commented that the complexity of modern life, difficulties of making a living, and technical nature of education were hindering parental involvement in schools. Sounds like déjà vu all over again one hundred years later doesn't it?

Prominent social critics begged parents to get more involved with their children's education. "Home and school leagues" were formed in towns and cities across the country to foster better communication and became commonplace in the early 1900s. By the mid 1920s there were strong state PTAs across the country.

With this rising organization of concerned parents came a rising level of frustration among educators. Parents were encouraged to get together socially but confine themselves to fund raising rather than getting too involved with education itself. Psychologists and social scientists had begun to study child-rearing practices and felt the need to teach parents how to parent. Parent's Magazine was founded in 1929 and the PTA came to be viewed as a vehicle for educating parents - not partnering with them.

After World War I, education expanded in size and scope. High school enrollments increased and junior high schools, ability grouping, and special education became popular. Parents were encouraged to take more responsibility for their children's education, to study available schools and make the best choice of educational options. Parents at the time were also encouraged to listen to, not dominate, their children, and to work cooperatively with the school.

"Progressive Education" Takes The Stage

By the 1930s, progressive education was in vogue, stressing the education of the "whole child". Families in the depths of the depression relied on schools for more than just education: they were looking for an answer to how their children could renew the American dream. Those with means were encouraged to take responsibility for "other people's children" in order to raise America up.

As immigrants and the rural poor poured into America's urban areas in search of work, educators were faced with the challenges of educating their children. School nurses, social workers, and guidance counselors joined the school staff. Children were tested and sorted and their psychological wellbeing was addressed along with teaching the multiplication tables. Schools tried to be teacher, parent, nurse, and confidant to a tired and hungry mass of children. Parents were too busy trying to keep a roof over their heads and some soup in the pot to offer much help.

Dr. Spock, what should I do?

After the challenges of the Second World War, Americans wanted to renew their commitment to and trust in American education. Dr. Spock was teaching them a warmer "child-centered" parenting approach and they wanted to indulge their children with the best of everything. Parents' roles were to raise confident, moral, happy children. The teacher's role was to educate them. The dreaded "parent-teacher conference" became a vehicle for cooperation between home and school and stay-at-home moms volunteered to manage bake sales and shelve library books. Sputnik scared us into a revolution in math and science education and the space race began.

In the 1960s, segregated schools that were separate and clearly not equal were challenged. Jonathan Kozol[15] began his series of books on the crisis of urban education and the disparity of opportunity for poor and minority children. I still remember reading the response of one mother in a Chicago housing project to Kozol's question about how the project's children survived its terrible conditions. "There are no children here" was her sad response. Children were robbed of their childhood in such an inhumane system.

The federal government began to encourage more parental involvement in education by requiring parent advisory programs in Head Start, the anti-poverty pre-school program of the 60s, and other anti-poverty programs of the decade. Teachers joined unions and statewide walkouts occurred, making them a force to be reckoned with in educational decision making.

Parents Demand New Rights.

By the 1970s, parents' rights were the topic of discussion. Americans were in a questioning mood, questioning the government, questioning authority, questioning the school's decisions about their children. The Family Education Rights and Privacy Act[16] became federal law, assuring parental access to student records and preventing release of student information without parental consent. Parents' groups formed in the Bronx and Harlem to defend themselves from an educational system they felt blamed the victim for lack of educational attainment. Parents of handicapped children won landmark legislation providing a free and appropriate education for every child regardless of handicapping condition[17].

The 1980s brought a flurry of reports on the state of American education. James Comer[18] advocated bringing the community into the school. James Coleman[19] studied education and inequality and promoted more respect and communication between the home and school to build both communities and the schools that serve them. *A Nation At Risk*[20] frightened America about a "rising tide of mediocrity" in American education, and fostered three more decades of school restructuring, outcomes-based education, and decentralization.

Today we seem no more certain of how to re-build American education than in decades past. We still struggle with the home-school partnership. Teachers complain about parents who abdicate their role and parents complain about teachers who can't teach. No Child Left Behind[21] promised much and has indeed given us much: much testing, much stress, but how much gain as the rest of the world races ahead of us into the 21st century. We "race to the top" for funding but can't decide if we should teach content or process or agree on standards for what 4th grade math students should know.

... No Child Left Behind has given us much...much high-stakes testing that is!...

We try making schools smaller, making them "magnetic"[22], making them more open, and inviting parents in. High schools are asked to build relationships, make classes more relevant, and offer more rigorous classwork[23]. We set benchmarks for promotion and have the kids eat a good breakfast before the high-stakes test that will decide if they are promoted to the next grade. We buy computers (surely technology is the answer, right?), pay students to take advanced placement classes and come to school (yes, we pay kids to do what is best for them), identify every child with a health impairment or learning disability (10 – 15% of American children are identified as handicapped in some way), and "medicate to educate" (allergies, asthma, hyperactivity, inattention, anxiety, oppositional/defiant disorder, mood disorder, bipolar disorder, conduct disorder, and the list goes on).

A Pill To Wake Up...

Teachers complain about parents who "won't medicate" their unruly children and I see children taking so many medications they can't remember what they're all for. One fifth grader told me he takes a pill to wake up, a pill to calm down, a pill to go to sleep, and another one for – something or other he can't remember. He also complains of headaches, dizziness, nausea, wants to sleep after lunch, is irritable, and has an explosive temper. So what am I trying to help with here, the little boy with emotional problems or the side effects of all his medications?

As I sit in meetings, trying to understand a child's learning or adjustment problem and trying to plan an educational program that works, several conclusions are inescapable. Most teachers work incredibly hard to help every child learn. They try to find a different approach. They call home and send notes. They ask each other for ideas and try whatever might work. They juggle the role of teacher, counselor, nurse, disciplinarian, and social worker as they grade papers, stop fights, wipe noses, and attend meetings. I see them lugging home papers to grade and lessons to plan every night and lugging them back in every morning. They are incredibly bright people who could have had many career options with higher pay. They chose instead to teach your children and hold part-time jobs to supplement their income and provide for their own children. Their job is not an easy one.

Traveling Down The Pre-Prison Track

Unfortunately, some also short-change history and science to focus on the tested subjects of math and reading. They have students write and write before the state writing test – but not so much afterward. When they serve over 100 students every day they allow themselves to be satisfied if Johnny at least passes biology without acknowledging that Johnny is gifted and the "D" he got should have been an "A". To meet the goal of zero tolerance and equitable discipline, they allow students to travel quickly down the pre-prison track of suspension - to drop-out - to inmate. They deny the public relations nightmare of gang activity as hand signs flash down the middle school hallway.

Most parents too are doing their very best and are the key to every child's success A single dad with hands hardened from labor who didn't finish school checks his son's homework every night. He tells me that he doesn't want his son to work for hours in the hot son like he does. When he can't figure out his son's elementary school math homework, he calls his neighbor to come over to help.

A young mother pesters the school – and me - until her daughter's reading disability is discovered and she gets the help she needs. There is a family history of dyslexia and she's noticed her daughter can't sound out words, substitutes and skips words while reading, is printing letters incorrectly, and can't spell. The teacher feels it's a little early to refer for testing – maybe she's just immature and retention is the answer. Mom says no, something is wrong. Turns out mom is right.

It Would Be Easier If She Quit School

Immigrant parents bring their child to school on the first day with new clothes and pencils, eager to learn, but not speaking English, sometimes after years with no educational opportunities. A mother tells me through a translator that it would be easier for her if her daughter quit school. She could help out at home with younger siblings and their battles over homework would end. "Oh no", I'm thinking, "another dropout." Instead she says, "I won't let her quit. Education is the only way for her to make something of herself, get a good job, and not have to travel forty miles to a 12-hour-a-day job folding towels in a hot textile plant like I do". Mom's 2 am to 2 pm shift had just ended and she had hurried the forty miles back to the high school to attend our meeting and convince her daughter to stay in school.

Parents volunteer, helping other people's children as well as their own. They are secret pals at Christmas and buy extra school supplies to give away. Parents bring their handicapped children in wheelchairs and rush to school for every bump or earache. They take off work to come to meetings, sometimes dealing with employer complaints and docked pay.

Other Parents Are Missing In Action

Other parents are, unfortunately, missing in action. Their children raise themselves. They don't know their children's friends – or their whereabouts. Many work nights – or even days and nights – and have no time or energy left to help with homework. They agree to a meeting with the teacher and then don't come. They enable bad behavior and wonder why it continues. They blame the teacher, blame the test, and blame each other. They allow their children to spend hours plugged in to an electronic reality and only unplug them long enough to carpool them to music lessons, soccer practice, and after-school activities. They call their kids to dinner and everyone gets in the car! They don't read together, play together, or even *be* together as a family in today's fast-paced society. Perhaps that's why children no longer even seem to respond to human voice!

... You know you're a working woman when...you call your kids to dinner and they go get in the car!...

No Child Left Behind. Race To The Top. Medicate to educate. Parents missing in action. Settling for less than their best. What does all this mean for American education? How are we doing? Do we have the educational system we need to renew America and find and secure our future amid the fear and uncertainty of a global economy in turmoil?

...Dropping out of school in not just quitting on yourself, it's quitting on your country...

President Barack Obama, State of the Union, Feb. 2009

6

So, Where are We Today?

Are American Schools Up To The Challenge?

Increasingly we are faced with statistics that beg the question: "Are American schools up to the challenge of the 21st century?" Indeed the Strong American Schools[24] project, a campaign of the Eli and Edythe Broad Foundation and the Bill & Melinda Gates Foundation, had as its goal to bring this concern to the forefront of the 2008 presidential election. Their initiative, ED in '08[25], provided a wealth of information that should have made every American demand that education become an immediate national priority.

The following statistics should frighten us all:

- 6,000 students drop out of school *each day*.

- Nationally, 30% of ninth graders don't graduate high school in four years.

- Our high school graduation rate used to be the highest in the world - we are now 19th.

- In some states and cities, *half of all minority students don't graduate at all*.

- 2/3 of newly created jobs require some college or certainly advanced training beyond high school.

- The technical reading required of many blue collar jobs is as difficult as that of college level literature.

- We have dropped from first to fifth in the % of 25-34-year-olds with a college degree; if trends continue we will have fallen to 18th by 2019.

- 65% of US convicts are also high school dropouts.

- 9th grade English teachers say they spend one-third of their time re-teaching skills that students *should have learned in middle school.*

- About *70%* of US 8th graders are not proficient in reading and will not catch up.

- One in three college freshmen needs remedial coursework; of those attending community colleges, it's 3 out of 5 and fewer than 25% earn a degree in *eight* years. These remedial programs are estimated to cost $2 billion per year.

- Only 56% of college freshmen earn a degree after *six* years – one of the lowest rates in the world.

- About half of all applicants fail entry level tests of math and reading according to the International Brotherhood of Electrical Workers.

- On a 2003 test of Problem-Solving for Tomorrow's World for 15-year-olds, 24% of the American students could not solve the simplest problems such as using a map to plan a trip or coordinating schedules with others.

- Relative to other countries that surpass us educationally, US students come from better-educated, wealthier families and go to schools that are comparatively better financed.

- A spring 2006 survey of human resource officials by the Conference Board found that 72% rated recent hires as deficient in basic skills such as grammar and spelling.

When my own 22-year-old son interviewed for a technical position at a major research university, he was a little concerned about remembering all his advanced math and laser technology accurately. The interviewer, who was head of the research project, assured him: "Don't worry young man, you did fine. I interview young people every day who don't even know there are three feet in a yard!"

States vary a great deal with respect to graduation rates[26] (in parenthesis). If you live in:

- North Dakota (80.9%)
- New Jersey (83.3%),
- Vermont (82.3%),
- Iowa (80.2%),
- Wisconsin (81%)

at least you are only losing about 20% of your students before graduation each year. Surely the citizens of your state will be willing to support those lost students *throughout their lifetime,* right?

However if you live in:

- South Carolina (54.9%),
- Georgia (57.8%),
- Nevada (41.8)
- New Mexico (54.9)
- Louisiana (57.4)

- North Carolina (57.8%) or
- The nation's capital, Washington DC, (59.5%)

are you really prepared to manage this huge population of undereducated, probably unemployable, and potentially angry young people? Will your citizens pay for their health care, incarcerate those who turn to crime, house those with mental illness and substance abuse problems, support their children, and provide for them as they age? Other states with graduation rates below 70% include: Washington, Arizona, Arkansas, Virginia, Texas, Alaska, Tennessee, Delaware, Hawaii, California, Mississippi, Alabama, and Florida. What is the cost to these states of statistics like these? Can you spell "a s t r o n o m i c a l"?

What Jobs Can We Offer The Nation's 9th Grade Dropouts?

Apparently not much has changed since the Graduation Project 2007 Diplomas Count report from Education Week[27], showed that 1.23 million students, most from minority groups, failed to graduate in 2006-2007. Typically more than 1/3 of dropouts fail to make the transition from 9th to 10th grade. There is tremendous variation across the country however. In some states *more than half the dropouts never make it out of the freshman class!* How will these children compete for jobs with less than a 9th grade education? How will they support their families and defend democracy? These low graduation rates have remained steady for years, as has the gender gap in which males, especially black males, are frighteningly unlikely to graduate. Do you see the relationship between *education* and *incarceration*?

The numbers are even more frightening if we look at recent graduation rates in America's large, urban school districts. Consider the graduation rates of these American cities:

- Montgomery County, Maryland graduates just over 83.1% of its students in four years
- Raleigh/Wake County, NC graduates 64.6%,
- Memphis graduates 61.7%,
- Houston 42.2%,
- Philadelphia 48.4%,
- Los Angeles 40.6%
- New York City 54.8%,
- Chicago 55.4%
- Clark Co. Nevada 39.9%!

Research compiled by "ED in '08" suggested that in fact improving graduation rates would reduce the costs of health care, crime, and poverty in America by perhaps 200 billion dollars annually. That 200 billion could be converted to retool the American engine of innovation and leadership, creating jobs for the next "American Century".

... And what could you do with $200 billion?...

What are the costs of this undereducated populace? A dropout is eight times as likely to be in jail or prison as a high school graduate. Compare the costs of educating a student - $9,644 – to the costs of incarceration - $22,600. Based on these figures, if we increased the high school graduation rate by even 1% we could save $1.4 billion *per year* in reduced costs of crime.

How To Contain Health Care Costs? Education!

Health costs are tied to education levels as well and certainly health care costs are uppermost in the minds of many Americans. High school graduates live over nine years longer than dropouts, while the college-educated are twice as likely to report themselves to be in good health.

Indeed common sense would suggest that the more educated an individual, the more likely they would be to want and be able to adopt a healthy lifestyle. In fact dropouts from just the class of 2006 cost us an additional $17 billion in health care costs. A few billion here, a few billion there, it starts to add up!

High school dropouts are also 25 times more likely to be on Medicaid, costing states $8,000 per dropout per year. Perhaps your family could contribute the $8,000 *each year* to pay the costs for one of these students!

What is the lost tax revenue of each year's dropouts and "off-shored" jobs?

Could you survive on $16,000 per year in income? Could you support your family? That is the average income of a high school dropout compared to $26,000 for a high school graduate – not exactly a living wage either. Still the $16,000 figure can sound like a fortune to an immigrant child from an impoverished background. Dropout prevention posters touting the $16,000 figure are clearly not a deterrent to such teens. Not only will all those dropouts require taxpayer support, they will pay perhaps $60,000 less in taxes during their lifetime. America loses $192 billion in combined income and tax revenue on each cohort of 18-year-olds who drop out of school – not to mention the lost tax revenue as jobs move overseas in search of more qualified and cheaper labor.

Again, what could America do with $192 billion or more to restore the American dream? Isn't saving billions and billions of dollars of welfare costs and lost tax revenue worth the nation's attention? Instead of waiting for government "bail outs" shouldn't we bail ourselves out by educating our children to ensure their future? Shouldn't education become "Priority One"?

... What could America do if we could recoup the lost tax income from each year's dropouts and outsourced jobs?...

If such statistics make your eyes glaze over, perhaps hearing about some more of the real students behind the numbers would move you to action. These are the students I see daily as a school psychologist in the public schools of America. They are the real reason that education must become America's national priority. While the names are fictitious – the students and their problems are not.

Donny, Gina, Brian, and Maria

Donny is a handsome, polite, and brooding young African American teen who has been in special education most of his school career. He tells me he's doing fine and doesn't need special education help any more. I ask him what he wants to do for a career. Engineering. I ask him how many credits he has for graduation (he is 18 years old and still a sophomore). He doesn't know. I count them up for him – he has five. I ask about his many suspensions from school. He shrugs and says he is sometimes late to class or uses his cell phone at school. He gets suspended for being tardy.

...If you are late to school today – don't come at all tomorrow. What's wrong with that picture?...

We complete the testing and I see that Donny is gifted – yes gifted – as in incredibly smart. I ask him if he realizes how bright he is – that he is clearly intellectually capable of becoming an engineer – if he could only graduate from high school! He shrugs. He's never even seen a college campus.

A week later, driving by the high school, I see Donny walking down the road – away from school and away from his future – cell phone in hand.

Gina is a bouncy teen who can't seem to pass algebra – after trying three times. I give her a math calculation test – she asks for a calculator as all my high school students do – and I tell her she has to do the problems with only scrap paper and her own brain! She asks if she can skip the problems she can't do without a calculator. I say yes and she skips long division, fractions, and decimals! I have other students who have supposedly passed Algebra but miss problems like -5 x 8. How can you pass Algebra and miss -5 x 8? If you can't do long division without a calculator, how do you know when to use division to solve a problem? I see high school students adding on their fingers! How do you get to high school adding on your fingers?

Brian is a fifth grader who is the class clown. He doesn't do his homework, doesn't finish classwork, and disrupts others. His teacher thinks he may be fairly bright and wonders how he can be failing. She convinces mom to agree to an evaluation of his ability and achievement. It quickly becomes clear that Brian can't decode words. He is actually very bright but is in danger of failing fifth grade. He admits he can't keep up in class, hates reading, and wonders how the "smart" kids do it. He doesn't even realize he *is* one of the smart kids – he just has a reading disability that no one recognized.

Maria is a young, dark-haired sixth grader the school staff feels is intellectually disabled. She has been in this country for some time but does not seem able to learn English and is failing all her school subjects. She has failed one grade already and is in danger of failing again, giving her an excellent chance of becoming a dropout statistic as soon as she turns 16. Because she does not speak English well, I use a non-verbal intelligence test on which Maria scores in the *well-above average* range. Here is a bright young lady, speaking only Spanish to her Latina friends at school, not understanding much of what is said in the English-speaking classroom, and on track to become a statistic. What a waste of her potential.

These are real stories of real children. Multiply them by thousands, perhaps hundreds of thousands, maybe millions, across America and we have a sea of lost opportunities that America can not afford.

...All who have meditated on the art of governing mankind have been convinced that the fate of empires depends on the education of youth...

Aristotle

7

Education Is a Process... Not a Place

Why aren't American children building the skills they'll need to survive in the 21st century? How and where are our children being educated today? We hear politicians debating school choice but do parents really have a choice? Typically, the only "choice" for parents is among their local public schools, some of which may provide a better educational opportunity for their child than others. The federal No Child Left Behind Act allowed parents to move their child from a "failing" school to another more successful school within the district yet few parents took advantage of the opportunity. Some districts offer magnet schools with specialized offerings or curriculum programs such as technology, international studies, or the arts. These programs are often offered in under-enrolled urban schools in poverty areas to attract more affluent families, mitigating the effect of concentrated poverty. Often slots at such schools are limited however, and won by the luck of the draw.

Some choice advocates go so far as to promote a choice among all educational options including private and charter schools. Indeed the federal government has actively promoted charter school options through funding priorities in the Race To The Money (RTT$), oops, I mean Race to the Top (RttT), competition.

Homeschooling parents would like a tax credit equal to the cost of the public school education they are rejecting. Dissenters cry that such changes in policy would cause an exodus from public schools that would drain them of talent and leave only the least able and most difficult to teach. Do parents – at least those unable to pay for private education – really have much of a choice?

The Right And The Left Are Both Wrong On School Choice!

Those on the political right correctly decry the nation's dismal educational outcomes but offer only choice and increased testing for accountability to fix the problem. Those on the political left insist that only universal pre-school and more money for innovation in public schools is the answer. Pundits and preachers on the right tell parents to pull their children out of those "anti-Christian government" schools. Pundits on the left want every child in preschool by age three and a government program from then on and speak derisively about the "anti-public school" crowd. The right and the left are both wrong! The problems in American education can't be solved by changing schools, abandoning public schools, increased testing, more money, or allowing a pre-school teacher to raise your child.

And where is the media? Have they investigated this national crisis and prodded the nation to action as they have rightly done in other areas of public affairs? Maybe now they will with the attention generated by "Waiting For Superman"[28], a book and movie which highlight the true crisis of education in America. With the likes of Oprah, Bill Gates, Michelle Rhee, Geoffrey Canada and Jay Mathews weighing in, we see why students and their families endure an agonizing wait for a few slots in local charter schools. We watch as other children and their families hope in vein, as did the Harlem

Children's Center's Geoffry Canada, for a superhero, a.k.a. "Superman", to swoop in and rescue them from a life of missed opportunities. Did you ever think maybe superman is us?

Do we realize that although we are all still waiting for some "superman" to deliver on the promise of high-skill/high-pay jobs, we won't have enough qualified Americans to fill them anyway.

The media typically prefers to focus our attention on the hot-button issues of parents banning a book with two daddies or pulling their child out of sex ed class. The "mainstream" media send in teams of news trucks to cover the horrible but rare cases of student violence. The "new media" write those teaser headlines about teachers having sex with students or a public school's refusal to use the word "Christmas". These headline-grabbing story leads, scary as they are, have very little to do with solving the real and pervasive problems in American education that affect millions of children every day. Shame on the media for not showing us that...

> *... the right and left are both*
> *...wrong on education!...*

So if the left and right are wrong on public schools, what is going on behind the school house door? Are public schools still even a viable option for educating America's children?

Public Schools Must Be Part Of The Solution

Clearly public schools serve the vast majority of American children - nearly 50 million of them – and always will. It is clear that no call from the religious right to "pull your children out" will dramatically change the fact that virtually all students are and will continue to be educated in America's public schools. These public schools are full of dedicated teachers who work hard every day, most evenings, and most weekends trying to educate your child.

Picture the kindergarten teacher with twenty or so young students aged not quite five to almost six. Some of these students have never had a remotely academic experience – perhaps having the TV as baby sitter. They have no concept of print, i.e. books tell stories, you read from left to right, letters make words, etc. They have never had to listen, sit still, eat a complete meal, respond to their name, or walk in a line.

Other children in the same classroom have had an academic pre-school, traveled, already learned their letters and numbers, and have mom and dad at home to read to them every night. The teacher's task is to shepherd all twenty through an increasingly academic curriculum that requires five-year-olds to read and write before they ever hit first grade. Yes, if your child isn't reading and writing little stories by the end of kindergarten, expect to be asked to consider retention!

A Tale of Two Kindergarteners

Two little kindergarten boys make the point of both the challenges and the successes of public education.

Jose was referred by his teacher not long after kindergarten began. True, he didn't speak English but he didn't even respond to his name – in fact he wouldn't say anything at all, not even in Spanish. He couldn't color, didn't come to circle time when called, couldn't use a scissors – he was barely at a pre-school level. Compared to other Hispanic peers, he seemed severely disabled. What to do? The speech therapist, occupational therapist, and I all observed him. A non-verbal screening of his ability showed his potential yet he made little progress. The teacher assistant spent extra time and asked for more ideas. The teacher was patient and supportive, trying different materials and approaches. A translator helped with parent conferences. Parent volunteers worked with him. Gradually we saw the spark of understanding. By the end of the year Jose was responding. He began to learn.

Spencer was referred by the end of his kindergarten year. The teacher felt there must be an intellectual disability as she had never had a child who had learned so few letters or numbers after months of trying. He still couldn't name common animals or count objects yet he chattered constantly and seemed to be trying desperately to keep up in class. As we walked to my office you could see the sparkle in his eyes as he rambled on about sharks and trucks and lunch and soccer as I struggled to keep his train of thought on the track! He told me about training his "pet shark" to roll over and fetch. I noted it must be hard for a shark to do such things. He thought for a moment and responded: "It's a dog shark!" That was my first clue there was no intellectual disability! A range of evaluations was completed by me, the speech/language specialist, occupational therapist, and school nurse. As we shared our impressions we identified not an intellectual disability but a language disability. Spencer was a very smart little boy who was struggling to match concepts with the words that describe them and to verbalize his thoughts. With special instruction he began to blossom.

Both little boys are lucky they are in public school where trained and caring people were able to help them. Private schools were not only out of reach financially for these families but are not often equipped to meet children's special needs. Homeschooling would have provided the love – but not the expertise – needed to help these little guys succeed. Public schools are the best learning place for Jose and Spencer.

Still public schools expect much, some say too much, of five-year-olds in kindergarten. Are these little children really developmentally ready to sit quietly, listen, write in the lines, and learn what used to be taught in first grade? Earlier in my career we were concerned about a child who hadn't learned to recognize all her letters at the end of kindergarten. Now we are concerned if she hasn't learned all the letters and their sounds by December. Kindergarten used to be the place where you learned to write your name. Now teachers come to me in September to say "he can't even write his name!" We seem to want five-year-olds to have the attention span of ten-year-olds and suggest a doctor be consulted about attention deficits and hyperactivity when they don't.

Remember Junior High?

Well, now it's called middle school and spans grades six to eight. It's where childhood meets adolescence in a head-on collision. Girls who still have Barbies at home and boys with high voices meet girls with children of their own and boys who shave. What's a teacher to do, especially when this diverse group is seated – or perhaps milling around – in one classroom?

Middle school associations and researchers across America have studied this pre-adolescent group and developed curriculum and programs directed to the strengths and challenges of pre-teens. Many middle schools do a wonderful job of guiding young sixth graders fresh out of elementary school, building the core skills of seventh graders, and preparing eighth graders to enter high school. It's exciting to see the happy chatter as students learn to remember their locker combination, run for their first school office, join their first club, and navigate the change of classes and requirements of subject-area teachers. I see them delving into history, hunched over a science experiment, raising their hand to solve an equation, and exploring the ever-changing social scene.

I am also told of the gang activity of a 16-year-old middle-schooler I am going to evaluate. He has a probation officer, walks with a confident swagger, and is as polite as can be when we meet. A new program has been developed to place such over-age "trouble makers" in high school to reduce their influence on the younger students. Not surprisingly, the receiving high school is not too happy with the program. Other students struggle with the demands of the middle school environment. They are missing basic math calculation skills, have little basic knowledge of science or history, and may still be struggling to comprehend as they read. Their backpack is a black hole, as is their locker. Homework? "I know I did it but I can't find it." "The project is due today? I forgot to write it in my planner." Young ladies dress like their idols on TV and in the movies, trying to look sexy and vying for attention. Other children are overweight, tired and listless. Their favorite activity? Playing video games. Favorite book? Favorite what??

Letter Sweater Or Trench Coat?

Many high schools bustle with the same energy and excitement most of us remember from our teen years. Some offer the International Baccalaureate program of demanding coursework. Most offer advanced placement courses providing college credit to those who work hard. Computer labs are filled with students who are called on to fix their teachers' and parents' computers. Students gain millions of dollars in scholarships, admission to top universities, go with their bandmates to parades and contests, and win science championships and football homecoming games.

Now picture the high school biology teacher with 100 students in the course of a day. Some could learn the material just by reading the book. Some can't read at all. Some are already alienated and angry – caught up in a teen subculture of rebellion. Some are dressed in black with body piercing and spray-painted hair. Others are "preppy" types whose parents will call about a "B" on a quiz. Some are so quiet you don't know they are there. Others can't seem to sit still or stop talking. Others have parents who expect an "A" in every subject for the all –important GPA and an increasingly hard-to-get spot in the college of choice. Others don't seem to have any parents at all.

But if we just lowered class sizes…, had more money…

Public school advocates complain of large class sizes, lack of funds, and disinterested parents. Still, the *Data 360* [29] report shows that the teacher/student ratio has actually decreased from about one teacher to 22 students in 1970 to about one teacher to 16-17 students today. Per pupil spending in constant 2004-2005 dollars has also risen from nearly $4,500 in 1970 to just over $9,000 in 2002. We seem to be spending more without attaining the improved outcomes we seek.

… Class sizes have gone down while per pupil spending has gone up. Are we getting the expected benefit?

Public Schools Don't Get A "Choice"

Remember that public schools are an institution that takes every child in the attendance area. No screening for ability, motivation, career interest, or level of parental involvement. As a public institution, schools are a microcosm of society – its strengths and weaknesses.

There are mostly great teachers and administrators – and some who are not up to the task. There are classrooms with the latest technology and materials and others without enough books to go around. There are bright, new schools where children walk in rows through sunny atriums to the cafeteria and others with "no-go" areas and bathrooms where bullies harass the weak. There are exciting projects, beautiful art, string quartets practicing, and students who lead service projects that impact people's lives.

There are also children who can't read, write a complete sentence, or solve a math problem. There are concerned parents who will come to school at a moment's notice and those who have a "you deal with him" attitude and a disconnected phone.

Clearly much good, hard, work is being accomplished for the millions of children in America's schools. As have generations before, students draw silhouettes of Abraham Lincoln every February, play tag on the playground, cheer their team on Friday night, carry in their carefully created science projects, and solve geometry proofs. It's still exciting to see the fresh faces on the first day of school, the decorated bulletin boards, Friday folders, and Key Club meeting announcements in schools across the country. It's what keeps those of us who work in the public schools coming back each year.

We know what we have accomplished, and we believe we can do it again. We withstand the latest "new thing" and keep trying to educate every child against every odd. We have always had poor children, immigrant children, handicapped children, and parent-less children and we still do today. But still we try to take each one and move them forward so that each can realize his or her own dreams. There is hardly a nobler goal or profession but still children, by the millions, are being "left behind".

...History is a race between education and catastrophe...

H. G. Wells

8

Other "Places" For The Process of Education

Since education is a *process*, not a *place*, there are other *places* where America's children can be educated.

Private Schools

Private schools are certainly nothing new, having served the more privileged members of society for generations. Today more than 6 million children attend private schools. There are a variety of private school options including non-sectarian, Catholic, other Christian, Montessori, etc. Private schools were in existence in America 230 years before Horace Mann helped develop public schools in 1837 according to Citizens for Excellence in Education[30], a group promoting alternatives to public schooling. Indeed in the past, Education Week has reported that many public school teachers in urban school districts actually chose private education for their own children.

Indeed my own children spent their middle school years in a small, private, school. They studied Latin and Greek, art and music history, and classical literature along with math, science, and history. They learned about the foundations of America's government by studying the Greeks and Romans. They know names like Herodotus, Plato, Caesar, Dante, and Chaucer. They understand that the middle east crisis has been going on for several thousand years and can interpret today's news from that perspective. The same can likely not be said for most children in America's public schools.

Approximately 81% of the nation's private schools have a religious orientation. Among the students in those schools however, not all belong to the specific religious affiliation. Indeed it has often been the case that non-Catholic families would send their children to Catholic schools in their quest for a quality education outside the public system. Even in the 1980s, the Coleman report lauded private schools for producing better cognitive outcomes, better character and personality development, safer and more orderly environments, and smaller class sizes. Of course one can clearly argue that the students whose parents can and do send them to private schools have the type of privileged background that would result in better academic outcomes anyway.

... Don't students in private schools start with an advantage of parental resources?...

Nationally, over 6 million students attend more than 33,000 private schools in America equaling about 11% of the U.S. student population[31] according to the Council for American Private Education. Most serve fewer than 300 students. Schools can be further subdivided by orientation as of 2007-2008:

Catholic	42.5%
Nonsectarian	19.4%
Conservative Christian	15.2%
Baptist	5.5%
Lutheran	3.7%
Jewish	4.7%
Episcopal	2.1%
Seventh-day Adventist	1.1%
Calvinist	0.6%
Friends	0.4%

There are also all-girl's schools, all-boy's schools, military schools, boarding schools, and schools for students with special needs such as learning disabilities, ADHD, substance abuse problems, or autism.

The average tuition ranges from $4,994 at Catholic elementary schools to over $27,000 at non-sectarian high schools. Still most children of wealthy Americans – some 80% of them – still choose public schools for their children. Recent public polls suggest that the public does generally believe that private schools provide a better education and that parents should have the opportunity to choose their child's school.

According to the 2003 National Assessment of Educational Progress, private students in grades four and eight outperform public school students in both reading and math[32]. In fact while only around 30% of public school students score as proficient in reading, half of private school students do. Writing scores also show a private school advantage, but only 40% of private and 22% of public school seniors appear proficient. History, geography, and civics scores are even lower but private school students continue to outperform. Still only 17% of private school seniors and 11% of their public school peers are proficient in history! What was that again about "doomed to repeat..."?

... Only 11% of public school seniors test "proficient" in history...

School climate has also been compared between public and private schools. In some areas, such as use of alcohol, drug use, and tardiness, there is perhaps surprisingly little difference. Public school teachers are much more likely however to report student disrespect, absenteeism, students who are unprepared to learn, lack of parental involvement, lack of parent support, and student apathy. Clearly these are areas where self-selection – who goes to private schools and who goes to public – makes a huge difference.

As far as threats and attacks are concerned, again there are not large differences. Even in private schools small percentages of students have been the target of bullying, hate speech, and other acts of victimization. From 4% - 12% of public school students report such concerns. One area where there is a vast difference however is the perceived presence of gangs. A full 25% of public school students reported the presence of gangs around or in their school according to the Dec. 2006 Bureau of Justice Statistics report[33].

...25% of public school students see the presence of gangs...

While many of the statistics on private school students seem related to their clientele, e.g. higher graduation rates, higher college graduation rates, more demanding graduation requirements, etc., one statistic is significant. Students in the lowest 25% of socioeconomic factors who attended private schools in 8th grade were nearly four times as likely to earn a bachelor's degree as similarly disadvantaged students who attend public schools. Hmmmm.

The Council for American Private Education makes the case that private schools are good for students, good for families, and good for America. They point to higher test scores, higher graduation rates, safer schools, as well as a focus on values. Indeed a 1999-2000 survey on Schools and Staffing looked at how private and public school principals ranked their goals for education[34]. While 80% of public school administrators listed "basic literacy" as a top goal, private school administrators listed "academic excellence" as the leading point of effort. Only half listed "basic literacy" as an important goal – possibly because private school students have already attained basic literacy. Not surprisingly private schools listed "religious/spiritual life" as important while that factor did not appear on the public school principal's wish list at all. A 2007 study by Professor William Jeynes showed a narrower achievement gap between races and economic backgrounds in religious versus public schools[31]. Indeed he found *no* gap among African American, Latino, and white students in religious schools when their nuclear families were intact.

Indeed many Christian schools aim specifically at instilling a "Christian world view" in their students. While religion has been all but eliminated in public schools, private school teachers and their parents are free to educate their students by "teaching the subjects of the Christian school in such a way that students develop a biblical worldview out of which to think and act"[35]. While some may cringe at such an approach, for the devoutly Christian parent a private school education provides a level of comfort that the family's beliefs will be not only tolerated but celebrated. Some Moslem families in America are seeking schools with an "Islamic World View" as well. Jewish parents and those of the Eastern religions seem more content to transmit their religious values outside the school setting. What world view is your child developing? Yours or someone else's?

Charter Schools

Charter schools are a relatively new and rapidly growing phenomenon. They receive public funds but operate largely outside the structure and accountability requirements of public education. They are typically created to provide a specific curriculum program or come from a specific educational philosophy and have recently found favor in Washington. They must develop a proposal and, if successful, are granted a charter from the state educational agency. Short of participating in required end-of-year testing, they may operate largely as they choose. States "racing to the top" for federal funds had to assure a level of school choice that embraced the charter school concept – a requirement viewed by some as an intrusion of "The White House" into the "State House". Twelve states passed laws leading to more children having access to quality charter schools. Several promising charter franchises have been developed including "Uncommon Schools" in Newark, New York, and Boston which had 98% of their grade 3-8 students score advanced or proficient on the 2009 New York State Math exams and 89% score proficient in English/Language Arts. The organization is trying to double the number of schools it serves in the coming years[36]. "Green Dot", which began working cooperatively with teacher's unions in California, has graduated 80% of its entering 9th graders within four years compared to the district rate of 47%. 76% of those students have been admitted to four-year colleges. Philadelphia's "Mastery Charter Schools" has raised student achievement dramatically with every member of the class of '09 admitted to college. "KIPP" schools have shown greater gains with the neediest students as well.

Over 1,000,000 American students attend more than 3500 charter schools in 40 states, the District of Columbia, and Puerto Rico[37] Currently it is estimated that 365,000 students are waiting for charter school openings nationwide, an average of 238 students per school. In my own state parents continue to call for more charter school options while the state legislature adopts a more cautious – some might say far too cautious – approach of limiting charter school opportunities. In their winning Race To The Top bid, they offered "charter-lite" in which failing schools can become charters within their current school district. Somehow that feels like cheating!

Charter schools are typically smaller than traditional public schools and serve a disproportionate and increasing number of poor and minority students – apparently those parents are seeking better opportunities for their children than they feel they can receive in public schools – a position that early data appears to bear out. A December 2004 Harvard University study showed that charter school students are more likely to be proficient in reading and math with greater gains posted among minority and low-income students[38]. Charter schools that have been operating the longest show the highest achievement gains. These achievement scores are of special interest in that charter schools indeed serve large numbers of poor and disadvantaged students – the very students public schools are worried they will be left with if parental choice is expanded. More recent evaluations are underway.

The US Charter Schools Organization explains the perceived benefits of charter schools:

- more opportunities for quality learning
- the provision of choice for parents
- more accountability in public education
- innovative educational practices
- new professional opportunities for teachers
- increased community and parent involvement

Parents and teachers choose charter schools for high academic standards, smaller class size, innovative approaches, or for their educational philosophy. With the overall average size of charter schools at 250 students – many public high schools serve 2000 or more – many parents seek out charter schools to provide their child with a more personalized education. Clearly not all charters are successful; many have had their charters revoked. Stubborn refusal to fund replication of successful models however is telling many students that their minds really don't matter.

Home Schools

Homeschooling, once the perceived purview of religious fanatics or other anti-establishments types, has gained popularity and joined the ranks of mainstream education. Today between nearly 3% of students in America are home-schooled, returning to a rich tradition of home education that was common in early America. Many families choose home schooling to provide a more religious educational experience. Others prefer the flexible schedule of home schooling, want to allow their child to pursue a special skill or hobby, or have a special needs child they feel better prepared to teach at home. Some see public educational curricula as too limiting, too focused on the content of gateway tests. Indeed a National Center on Education Statistics report notes that 85% of home schooling parents list "dissatisfaction with the public school environment" as a reason for their decision. "Religious or moral development" was listed by 72%, and "dissatisfaction with their child's academic instruction" was listed by 68%[39].

Many of the preconceived notions about home-schooled students are being dispelled by the data as well An analysis of the studies, though relatively few in number, reported in "The Old Schoolhouse" magazine consistently shows home-schooled students performing quite well on standardized tests including the SAT. In fact two major studies carried out by the National Home Education Research Institute and the well-known ERIC Clearinghouse on Assessment and Evaluation at the University of Maryland confirm that home-schooled students do quite well academically compared to their publicly- and privately-schooled peers[40]. It is also interesting that no differences in achievement were apparent based on the homeschooling parent's level of education, the type of curriculum used, or the level of oversight by state agencies – all areas states often insist on regulating for homeschoolers. National spelling and geography bee champions have been homeschoolers in recent years, as has Heisman Trophy winner and NFL draft pick and education advocate Tim Tebow.

More and more colleges actively recruit home-schooled students, providing specific application information to prospective students from home-schooled families. These students consistently score well on standardized tests and tend to be organized self-starters who are well able to manage the college experience. Perhaps the names of the following colleges deemed "welcoming" to homeschoolers will be familiar to the reader:

Arizona State
Stanford
Johnson and Wales
Yale
The University of Delaware
Drexel
Southern Illinois University
Drake
The University of Iowa
Kansas State
Louisiana Tech
Towson State
Harvard
Northwestern
The University of Nebraska
Duke
Kent State
West Point
The Ohio State University
North Carolina State
Vanderbilt
The Art Institute of Pittsburgh
Hampton
Rhode Island School of Design

... But what about socialization?...

"But what about socialization?" everyone asks. Such a question probably amuses most home schooling families who are busy taking their children to piano lessons, community soccer league practice, and science classes at the local museum right along with their public school friends. Most communities have active home school associations that offer sports leagues, art contests, science fairs, and even proms and graduation ceremonies. Other families would question why their child should be socialized in public schools in which gang activity is present and students hesitate to use the bathrooms for fear of intimidation. While such problems are much more the exception than the rule in public schools, the criticism of the social development of the home-schooled child clearly seems misplaced.

Virtual Schools

A relatively new option that can work in any educational setting is the virtual or electronic classroom. With today's technology a student can be at home or at school, any time of day, studying advanced placement world history, astronomy, the history of American theater, or any of a wealth of courses.

Small high schools that would not have the resources to provide such classes can now offer them for the price of a computer and internet connection. Complete virtual schools have also sprung up to provide an entire year of coursework to home-schooled students or those who, for health reasons, are unable to attend school. States such as Florida and Michigan have created "Virtual High School" programs accessible to students across the country. The courses are accredited and designed and taught by educators. Colleges and universities now make online courses available to high school seniors as well.

21st Century Technology In A 20th Century High School

The magnet technology high school in which I worked as a high-school restructuring grant administrator experimented with "electronic electives" such as veterinary medicine and 20th century American music. While the courses piqued student interest and excited parents, there were challenges as well.

The school was on a modified year-round calendar that started at the end of July – well before the on-line courses were up and running for the fall semester. Students also needed a combination of motivation and perseverance to sit through 90-minutes a day of on-line instruction as part of the school's block schedule. Although the course instruction was totally on-line, the class still required a classroom instructor. No leaving twenty teens in a room by themselves for 90 minutes!

We also couldn't let students take the class at home. Students have to have a certain amount of "seat time" and 180 days of instruction to get credit for the course. Thus this exciting 21st innovation just didn't fit into a 20th century high school – even one focused on leadership and technology.

Private, charter, and home school parents may be able to make better use of virtual education. Several virtual schools market themselves through the Internet. Whether you are seeking a general curriculum, classical curriculum, a Catholic perspective, or a strongly college prep approach, there is likely a virtual program to meet your needs. Look for one that is accredited if you hope to pair courses with an eventual return to public schools. Accredited courses document the "seat time" and grades to be included in a public high school transcript. Virtual schools can also provide "credit recovery". If your child has failed a required course often it can be made up with an online option. This may be particularly important if your child is unable to attend public school summer school, or just needs a different approach. Make sure the virtual option will be accepted as part of your child's requirements however.

Clearly, education is a process...not a place.

Education can and does occur in public schools, private academies, Catholic parishes, across the Internet, and around the dining room table. The key for making education the national priority it should be lies with you – the parent – taking charge, informing yourself, and investing your time in the education of the children God has given you. You are truly the key to your child's success, and thus to America's Renaissance. As President Obama recently noted: "no program or policy can substitute for a parent" and the "responsibility for a child's education begins at home". Are you ready to take up the challenge?

9

But My Child's Public School Is OK - Isn't It?

Will Your Child Be Able To Get A Job?

As noted in the "ED in '08" policy primer, schools have always been the pathway to a better life for millions and millions of American children. Education opens the door to unlock potential and create opportunity. It is the great leveler of race, family income, and language. The American dream is in danger of becoming a nightmare for too many American children though. They will not be able to read well enough, write well enough, or "do the math" well enough to get a job.

As "ED in '08" reminds us, this is not somebody else's problem, it's our problem. Will the teen you see walking down the street in your neighborhood become a doctor or a drug dealer? Will your neighbor's children support your retirement or will you have to support them forever with your tax dollars? Will your daughter graduate deep in debt from college loans only to find out she has to compete for a job with better-prepared students from around the world?

Not too worried about your own children? They're in a good school in a great neighborhood, right? Well maybe not. The December 2007 edition of *The Education Reporter*[41] notes that in dozens of California neighborhoods where families spend many hundreds of thousands on their homes, few students score at proficiency in course courses such as Algebra,

Geometry, and English! They may sport a great looking GPA but how does that inflated "A" compare to the hard-earned "A" in other countries?

Are you a frog in hot water?

You live in a progressive state with a top-notch school system right? No? Well, your school district has great schools, right? Well, at least your kids go to one of the best schools *in* the district, right? Remember the story about the frog in the boiling water? If things get hot slowly enough, the frog never notices until it's too late! It's kind of like that with American education. We just assume everything is OK. "Bs" and "Cs" on the report card – not too bad I guess. I mean "B" is good and "C" is average, right? Doesn't really read too much but who does these days? Never seems to have any science homework but that will come later, right? The school seems pretty good I guess. We complain about education as a whole but we are so glad our own child's school is grade "A".

Grading America's Schools

The annual "Quality Counts" survey by Editorial Projects In Education Research Center and reported as an *Education Week* special report[42], reminds us that we are frogs – lulled into a false sense of security while the water gets hotter and hotter, eventually reaching the boiling point. By looking at a range of indicators, American education earns only a "C". Will a "C" be good enough to restore the American dream and keep us competitive with the rest of the world? Will "average" be good enough for an American Renaissance?

In the recent "Quality Counts 2010" report, states were given updated grades on various areas of performance: chance for success, standards, the teaching profession, and school finance.

The *"chance for success"* factor looks at the educational opportunities provided for the citizens of the state, with K-12 education being the most significant factor impacting career success. K-12 achievement is broken down by socio-economic status and other issues of equity that are so important to closing the achievement gap between white and black, rich and poor.

"Standards and accountability" refers to how well states set and measure educational expectations. In other words, what are students really supposed to know and be able to do, and how will we know they have mastered those skills? This survey reflects which states have signed on to the recently completed national standards in language arts and math.

The *"teaching profession performance"* area looks at not only whether teachers are well-prepared, but if they are supported, evaluated, and compensated effectively.

"School finance" evaluates school expenditures and equity of resources.

So America as a whole gets only a "C", but what about your state?

Well if "B" is "good", these states are good:

- Maryland 87.5
- New York 84
- Massachusetts 82.5
- Virginia 82.3
- Ohio 81.2
- Pennsylvania 80.5
- New Jersey 80.4
- Florida 80.3
- West Virginia 80.2

- Arkansas 79.9
- Vermont 79.5
- Georgia 79.5

Aren't you all glad we're using a 10-point scale and rounding up? In most schools everyone but Maryland would be taking home a "C".

Haven't seen your state listed yet? Uh-oh. Quality Counts gave you a "C" if you scored above a 70. Don't you wish your algebra teacher had been as generous?

- Texas 78.1
- Michigan 77.6
- Wisconsin 77.2
- Indiana 77
- Tennessee 76.9
- California 76.8
- Iowa 76.8
- Maine 76.5
- Delaware 76.5
- Oklahoma 76.4
- New Mexico 76.4
- Connecticut 76.4
- Hawaii 76.1
- The US 75.9 (Hmmm, that was a "D" when I was in school)
- Louisiana 75.8
- New Hampshire 75.6
- Minnesota 75.4
- Washington 75.4
- Alabama 75.3
- North Carolina 75.1
- Rhode Island 75
- Wyoming 74.7
- North Dakota 74.6

- Kansas 73.4
- Kentucky 73.3
- Utah 72.7
- Colorado 72.5
- Missouri 72.4
- Illinois 71.8
- Arkansas 71.3
- Oklahoma 71
- Idaho 70.9
- Montana 70.9
- Arizona 70.8
- Mississippi 70.0
- South Dakota 69.5 (thank goodness for rounding)

Still haven't seen your state? BIG Uh-Oh! Bringing up the rear of the 50 states with what in any school would really be an "F"...

- Nebraska 68.9
- Nevada 68.7
- Our nation's capital, Washington DC, 68.3

OK, so there are some bright spots. Massachusetts, New Jersey, New Hampshire, and Connecticut rated grades of "A" in the category of "Chance For Success", a measure of education's role in benefiting the people of the state. Many states got an "A" for adopting national Common Core achievement standards, South Carolina rated an "A" for teaching qualify and support, and Wyoming rated an "A" for the level and fairness of public school funding. Twenty states now link students' records – including their achievement – to teacher records. Only thirteen states actually tie the two together for evaluation.

When will these higher standards pay off?

Nineteen states now have an "A" in adopting and evaluating the more rigorous national standards. They would certainly argue that those new and stricter standards about what students need to learn should eventually yield the achievement results we should be demanding. We all desperately hope so! Still yet to come – and be paid for - is the curriculum, materials, teacher training, and assessment instruments to deliver on the promise of 21st century standards.

You may also be surprised at the difference in expectations and preparation of the teaching profession across the states. Almost all the states require a test of basic skills and subject knowledge for teacher certification, but few states assess the aspiring teacher's skills at *teaching* that subject matter or require that parents be notified if their child's teacher is not certified in the subject he or she is teaching.

Is Your Child's Physics Teacher A – Physics Teacher?

Did you assume your child's physics teacher is trained and certified to teach physics? Not necessarily. Your child's algebra teacher really is an *algebra* teacher, right? Not necessarily. There are classrooms where long-term subs fill in for weeks when no certified teacher can be found.

Racing Toward the Funding Cliff

If states are racing toward a deficit cliff as federal recovery funds dry up next year as many predict, which states have the farthest to fall? Wyoming had an "A" in 2010 for funds available for education and equitable distribution. States already failing (using my algebra teacher's grading scale!) included: North Dakota, South Dakota, Louisiana, Oklahoma, Texas, Arizona, North Carolina, Mississippi, Tennessee, Utah,

Nevada, and Idaho. What will these states do as they face perhaps a multi-billion dollar shortfall of state funds? What about growing districts like Wake County North Carolina where administrators face billions in school construction costs they can't afford that still may not even keep up with the enrollment increases they project for the next decade?

Of course for every statistic there is an explanation – some might say an excuse. As states review reports like these and the media comes calling (assuming the media does their job), states and politicians up for re-election will certainly want to focus on their improvement efforts. Clearly the lack of higher educational "results" can be considered a problem of "inputs" as well as a problem of "process" or policy.

While businesses can control the quality of their "inputs", schools cannot. American schools struggle to educate millions of children, many of whom have trouble learning and behaving. Over 100 different languages are spoken by American school children – but not American teachers. While a business can tweak a specific manufacturing process to improve productivity, tweaking the educational process to get a specific outcome is far more difficult.

So who is going to fix this? Who is responsible? Why doesn't somebody do something? The answer to "who" is – you and I! The answer to "why" is because we – you and I as parents – have abdicated our responsibility. We are the key to our own children's success. If we don't ensure our children receive a top-notch education, who will? If America's biggest businesses are investing in Asia's students, we must invest in our own. We are the key in our own house, in the school house, in the state house, and in the White House!

...there is no program or policy
that can substitute for a parent...

President Barack Obama, State of the Union Address,
Feb. 2009

10

Parents Are the Key: What Can I Do In My House?

Your house is THE most important house for your child's education. You hold the key to your child's expectations, their health, the environment in which they grow, what they watch on TV, who they associate with, whether they have oatmeal or a candy bar and a soda for breakfast, whether they have books to read – the list is endless. It's a big responsibility, and one no one else can do for you. There's no real substitute for mom and dad.

In their recent book, "I Just Want My Kids To Be Happy"[43], Aaron Cooper and Eric Keitel make the case that American parents are doing their children a disservice by emphasizing "happiness" at the expense of the qualities of character needed for true long-term success in life. They note that we parents are only too ready to say "yes" to everything, to buy anything, to make any excuse, so that our children will stay "happy".

They suggest instead that parents plant and nourish such seeds of long-term happiness in their children as good health, kindness, gratitude, optimism, natural talents, mastery of knowledge, and self-control. All these "seeds", if nourished, will carry over into your child's job from ages 5-18 (or so) – becoming educated. They will prepare your child for the tasks of education: attention, respect, listening, practice to mastery, diligence, responsibility – all the qualities needed for your child to learn in whatever "place" that learning is presented.

Education Priority 102

Whether your child must learn to listen in class or listen to you over a lesson at the kitchen table – your insistence that he does listen is the key. If you insist on practice to mastery, your child will learn that high goals are worth working for and that not everything – in fact almost nothing – comes easily. The work ethic and values you instill in your child today will help her grow into the woman you want her to become.

So what exactly should you be doing at your house? I have some ideas…

…Establishing lasting peace is the work of education; all politics can do is keep us out of war…

Maria Montessori

Be A Parent, Not A Pal, And Create A Family

- Not everyone with a child is a true "parent". Not every group of people who share living space can be called "family". You must BE the parent your child deserves and CREATE a loving, supportive, family in which your children can learn and grow.

- Insist on respectful listening. I am constantly amazed at children who continue to talk to each other when an adult is speaking to them. Are they so used to a constant stream of noise in their lives that your voice or that of their teacher is just one more annoyance?

- Turn it off, turn it off, turn it off. The TV that is. Why would you allow the TV to raise your children? Why would you allow their minds to be filled with violence, sexual images, advertising, and off-color humor? Don't abandon your children to grow up in the electronic village without you.

- Since it apparently needs to be said: Eat dinner together! The people who live in your house need to get to know each other! The dinner table used to be – and can be again – the place where the news of the day is shared and each person feels valued. It's where "what did you learn in school today" needs to get answered.

- Create family traditions based on your cultural, ethnic, and religious heritage. Such traditions have value to your children as they struggle to understand who they are and find their place in an increasingly global society.

- Help your children understand what you believe and why. How would your children answer questions like these? Every Sunday we..., My favorite holiday is...

because my family always ..., My mom always says...,
the most important thing is....

- Take a moment to consider what memories and values
  you are instilling in your children that will help them
  thrive and nurture their own families in the 21st
  century? What kind of parents do you think your
  children will be?

- Say "NO". It's OK, really! You don't have to take them
  everywhere, buy them everything, enroll them in
  everything, in order to keep them happy. I see the
  results of too many "Yes"s in children who have no
  limits, no boundaries, no willingness to stick with a
  hard task, no interest in anything that seems even
  remotely difficult.

- Don't enable bad behavior. Who hasn't written a tardy
  note for a teen who just didn't get out of bed in time?
  Who hasn't edited a paper for a child who had a soccer
  match the night before the due date? Who hasn't bailed
  out a teen – figuratively or literally – or provided more
  and more money to a spendthrift child. But we want
  them to be happy! We can't stand to see them fail.
  We're just helping him right? Yes, we're helping him
  avoid learning the responsible behavior of an adult-in-
  training.

- Teach your boys to be gentlemen and your girls to be
  ladies. Don't allow vulgar speech or disrespect. Don't
  buy your little girls sweat pants that say "cutie" across
  the rear!

- Teach "r2d2" – not the robot but "respect, responsibility,
  discipline, and determination"! I cringe when I see
  children throwing out a plate full of food after one bite

because it's "nasty", demanding a toy in the store, hitting a parent who says "no", continuing to interrupt their parent's conversation, rolling their eyes at a 30-minute reading requirement, and giving up when the answer isn't one word printed in bold in the science book. What kind of adults will such children be?

- Children should be *respectful*, of you, of others, of property, and of those in authority.

- Show them that *responsibility* means doing their share of family chores, saving money, doing their homework, being on time, and taking care of their belongings.

- Being *disciplined* doesn't have to mean being punished by a parent or teacher. The disciplined child shows restraint, doesn't waste, asks before taking, doesn't whine, doesn't need an immediate reward, and takes "no" for the answer.

- *Determination* overcomes challenges, keeps working when the task is difficult, practices hard, learns from mistakes, and takes pride in accomplishments.

Decide On A Process For Your Child's Education

- Yes, you decide. Remember that education is a process, not a place, and there are many options for your child. You must decide what is best.

- What about pre-school? Is your child better off at home with you? Probably, but perhaps that is not an option if you work. Do you choose daycare or a babysitter? Check and re-check references.

- Whatever is right and feasible for your family, remember that these years are critical ones for your growing child. They are developing trust in others, learning language, and studying the world around them. Talk to them. Read to them. Listen to music, play in the rain, and watch the sun set. You'll never regret the time spent introducing your child to the big, wide, world.

- If you feel your child's skills are not developing as you expect, talk to your doctor and insist that he or she listen to you! When you know something is wrong don't accept "she'll grow out of it" as a diagnosis. I have had many parents whose own Internet research set the stage for identifying the characteristics of autism or dyslexia and getting critical early intervention.

- Even when family members resist, follow your instincts and find someone who will listen and help. No one wants to find out their precious child has a disability but early intervention is key to a better outcome.

- Read to your child daily from an early age. That's how children learn the tools of language. Thoughts can be expressed in words and pictures. Letters have sounds and make words. You start at the front of the book and go to the back. Did you think such skills were innate? Ask any kindergarten teacher – they're definitely not.

- As kindergarten approaches, look at your child and consider what is important to you about his education. While virtually everyone will send their child to the public school down the street, you can choose a different path.

- Perhaps you will choose a private school of your religious affiliation. You may be interested in a Montessori approach. Perhaps your child has special needs, gifts, or talents that you are not sure can be adequately dealt with in a public school setting. What should you do?

- Visit the public school that your child would attend. Ask to visit a classroom. Talk to the special needs staff if your child has a disability. In my recent experience, children with special needs are well served by the expertise of the staff in public schools. This expertise is not found in most private schools and can not be fully met by the most loving and dedicated parent. Your public school provides occupational and physical therapy, behavior specialists, school nurses, psychologists, large-print books, adaptive physical education – all at no additional cost to you as a tax payer. A public school education is a hard-fought-for right for any handicapped child.

- In considering, or re-considering, an educational process for your older child, again visit the school your child would attend. Talk with other parents and children about the school and their experiences. The school's test scores, discipline data, etc. should be public information and are often posted on the district website.

- Consider your child's personality. Is he a follower or a leader? Does she have a special talent that needs to be nurtured? How will she fare in a middle school with 16-year-olds? (Yes, there are 16-year-olds in middle school).

- What about high school? Again, do your own research. Visit the school and get a feel for the climate. Is it clean and well maintained? What is the graduation rate? Will your child be able and encouraged to take advanced math and science? Visit an advanced math class. Do you see students in there who look like your child e.g. African American, female, Latino? If not, why not? What foreign languages are available? Does it feel like a place where your child would thrive or one in which he'd be lost in the crowd?

- Is private school a better option? Again you need to do the research before assuming that you get what you pay for! Certainly private school tuition will be a financial drain on your family but in some cases your child's school and life success may be at stake. You may also determine that e.g. a Catholic education is critical for your child or that an all-girl's school is what your daughter needs to succeed. Many private schools offer scholarships so don't assume that tuition is out of your reach.

- Home-schooling is not an oxymoron! To school your child at home has never been easier or more accepted. You can purchase a complete curriculum, enroll your child online, or share teaching duties with other homeschoolers. You can choose a curriculum which matches a Christian worldview or pick and choose materials based on your child's interests. There are also resources that explain the typical sequence of topics in each subject. Most communities have homeschooling groups that provide field trips, share materials, and offer other resources. For some families and some children, homeschooling is the best choice.

Whatever The Process, What Is The Key To Success?

- At the risk of being repetitive, turn it off, turn it off, turn it off! Turn off the TV, take off the headphones, shut down the video game. You are allowed to do that you know because YOU ARE THE PARENT! Your child is just that − a child. He or she is not old enough to make good decisions about what is important or how to spend his or her time.

- Save the electronic entertainment for a special treat or to view something − perish the thought − educational! Yes I know it seems hopeless but if you don't unplug your child's brain from borg-like cyberspace who will? Picture your children with all those wires coming out of their ears, speakers over their mouths, and lenses where their eyes used to be! When your little cyborg is plugged in she isn't thinking, dreaming, imagining, or creating, - just absorbing − who knows what - in the electronic village!

- Read to your children and have them read to you. Children learn to be better readers by listening to language. You won't know if they are learning to read unless you listen to them read to you.

- You don't have to buy a bunch of books. Get a library card and go there with your child. Let them see you reading. Talk about what you've read and ask them about what they've read.

- Create a traveling library of shared books and magazines in your neighborhood.

- If you feel your child is not reading simple picture story books fairly easily and fluently by the end of first grade, talk to the teacher.

- If your second grader is not reading and enjoying short stories on topics of interest, talk to the teacher.

- If your third grader is not able to read fluently from her science or history book (if they still have such things!) and talk about what she's read, talk to the teacher – even if her teacher is you. In my experience when the parent thinks something is wrong with their child's learning, it probably is.

- Make sure your child does assigned homework. If your older child has none, talk to the teacher since "no homework" is an unlikely scenario.

- If your child does not understand the homework, talk to the teacher. Ideally homework should be started in class as "guided practice" so the teacher can make sure each child understands the concepts to be practiced. If you are teaching the homework every night, something is wrong.

- If you are homeschooling, maintain a consistent daily and weekly schedule. It is easy for homeschoolers to think, "Oh, we can do that later", then "later" never comes!

- Be very realistic about your child's progress in homeschooling. Is he really making adequate progress? If not, don't wait. Seek assistance so that learning problems can be addressed.

- Review your child's writing skills – wherever they go to school. Can your second grader write several, short, complete, sentences on a topic?

- Is your elementary schooler learning cursive handwriting? Can she compose a short letter and address the envelope?

- By fourth grade can your child write a coherent, page-long story on a given topic?

- Is your middle-schooler writing logical and thorough essays and multi-page reports easily?

- Is your high schooler doing quality research and writing persuasive papers – not just "Google, cut, paste"?

- Find out what your child has learned about science. In many cases, it may not be much. He may not have a science book but may be doing science "projects". How can a high school student learn biology if he doesn't have a factual basis in life science in middle school and basic knowledge of animals, habitats, plants, and the human body from elementary school?

- What does your fifth grader know about the planets, electricity, weather, magnetism, animal habitats, the human body, how plants grow, the oceans, birds, and insects. You may be shocked at her lack of factual knowledge about the world around her.

- Check your child's knowledge of history and geography. Does he know as much about America and the world as you learned in school?

- Have a world map and globe in your home and look at them frequently. Save money by picking them up at garage sales. Watch the news together and locate the places mentioned.

- Can your older elementary student name the oceans, the continents, the states and capitals? What does he know about the ancient world? The middle ages? America's independence?

- Don't believe those who say there's no need to remember facts anymore because you can always look them up! Are our minds really supposed to be so empty of knowledge that everything must be relearned every time we need it?

- Set the expectation for your child to remember what he learns beyond the test. Are today's children really so much less capable of remembering learned knowledge? Do they understand that learning is more than a correct answer on the "multiple guess" exam?

- Your middle schooler is likely studying history "by hemisphere" – a somewhat awkward approach it seems to me. Again, ask her what she actually knows - what she can remember - about what she's learned. Look at her tests. Is she required to answer in complete sentences and compose essays or must she only use "multiple guess" or fill in the blanks? Perhaps that limited "sound bite" knowledge will come in handy in text messages some day but it's hardly enough to create an educated citizen!

- Do your children know about the great civilizations of the world? Do they understand how language, writing, law, and governments developed? Can they relate the world's wars and dictators to the freedoms they take for granted today?

- Read your child's history text. Yes, read it yourself. You may be shocked at the content (or lack thereof) and tone. Is that the history you remember?

- Read the stories in your child's reading book. Are you satisfied with the content? If not, you could try fighting with the school board. Some parents have done it and won but others have found out that parents have few rights with regard to curriculum.

- It may be necessary for you to supplement your child's curriculum. Yes, you may have to do just that! If your child's history book covers the civil war in three pages and the Great Depression in two you can find a wealth of interesting and accessible material on the internet or library.

- Create your own family history. Don't let your grandparents' stories of the depression, their experiences in the holocaust, or their fight for the right to vote go unrecorded. Don't let the struggles and triumphs of your ancestors be overlooked.

- Two of my children came home from high school to report errors in world history texts and with the teacher's explanation of some of the religious history and beliefs of our faith. My son researched his concerns and brought his history teacher several books to make his point. His teacher read the books, was surprised at what he hadn't learned in college, and agreed to change his presentation. My daughter shared her concern with her teacher who noted with apologies, "I'm really sorry. I never knew!"

- If your child's reader is filled with simplistic stories and boring characters you can provide classic literature at home. There is a wealth of interesting and exciting literature from some of the greatest authors of all time. Real adventure, mysteries, romance, history, fascinating characters – all waiting to stoke your child's imagination and love for literature.

- If you can, create a home library. Not only does that send a message to your children about what's important, it allows them easy access to timeless literature and topics in science and history. Yard sales are great for picking up books for very little investment. Expect 25-cent paperbacks and $1.00 for hardcover. Offer to share your finds for a classroom library.

- Instead of the usual babysitter, movie star, or TV show books, you can insist on quality works that have nourished children for years. If they aren't on the book report list – ask. It's unlikely your request for quality literature for a book report would be turned down.

- Search the internet for book lists of classic works by grade level, reading level, or historic era. Historical novels are a great way to combine reading and history.

- Check your child's math skills. Grade-appropriate math skills sheets are readily available online or from your local school supply store. Can your first grader add and subtract single numbers? Can she solve simple problems with picture clues?

- Can your second grader add and subtract with regrouping (you may remember that as borrowing and carrying!). Has he memorized basic math facts?

- Has your fourth grader memorized her multiplication tables and can she use math to solve simple problems around the house? Could she find the perimeter and area of her room? Does she know what time it will be five hours from now? Does she know which is the better bargain: 4 for $1.20 or 3 for $1.00?

- By middle school your child should be able to work easily with fractions and decimals, solve a multi-step problem and solve a proportion.

- Surely your highschooler does not need you to tell him how much a jacket costs that is on sale for 30% off, right? How about for an "early bird special" of 15% off the markdown price?

- Observe American holidays as a family. President's Day (remember when Washington and Lincoln had their own days?), Memorial Day, the 4th of July, Labor Day, Veteran's Day – all are important for children to understand.

- Ask your high school student what each holiday commemorates but don't be surprised if she has no idea. Those holidays don't merely represent a sale day at the mall! They commemorate important facets of American culture that should be shared with our nation's children.

- Make sure your child knows the Pledge of Allegiance and the words to the Star-Spangled Banner. Share the meaning and importance of each.

- Teach your children the American story through our music and art. As children we sang "America The Beautiful", "God Bless America", and studied American artists and authors. There are great books available on our American artists and musicians and Internet resources are almost limitless. Older students may enjoy learning about folk, gospel, jazz, the birth of the blues, American musical theater, the Harlem Renaissance, and (of course my favorite!) the roots of rock and roll.

- Make sure your child does not miss out on the opportunity to know how generations of Americans have viewed their country through art and music. Let them hear the songs of the south, railroad-building songs, Appalachian music, and certainly the protest songs of the 60s.

- Encourage your child to be respectful of other children's beliefs. This does not mean you must accept beliefs that differ from your own. It does mean than no child should be harassed or made to feel anxious or afraid at school. Tolerance means recognition and respect. Tolerance is also reciprocal.

- Teach your child the "soft skills" so necessary for success in life – those skills like manners, listening, being on time – that employers say are missing in today's would-be employees.

- "Please" and "thank-you" matter. Being on time each day – 8:00 am doesn't mean "sometime before 9:00" – is critical if you want to keep you job. I'm convinced that the teen who holds the door for me as I enter the high school in the morning will go much farther than the one who lets it slam in my face!

- Find time to listen to your children. Really listen. Are they trying to tell you they can't read, that they are being bullied at school, they're being pressured by peers, they can't see the board, or they're worried about the world or if dad's losing his job? Don't ignore them, yell at them, or dismiss them as unimportant. If you keep the door open, they will eventually come in.

- Know your children's friends and their families. A friend of mine commented that she would never let her child associate with a child whose family she didn't know. A quick phone call, "Hi, I'm Ashley's mom, and I just wanted to make sure you'll be home while they do homework" sets an expectation for Ashley, her friend and her friend's family. If your call seems unwelcome, your child doesn't go.

..The responsibility for a child's education begins at home...

President Barack Obama

Never doubt that a small group of thoughtful, committed people can change the world. Indeed it is the only thing that ever has.

Margaret Mead

11

In the School House

This Too Shall Pass...

Many educators take a "this too shall pass" approach to the latest education reform. Perhaps this is understandable as we sometimes seem to lurch from old math to new math, phonics to whole language, social promotion to high-stakes promotion tests, and back again. We ask for donations from business but don't want their suggestions since children aren't "widgets" on an assembly line. We complain that teachers have it easy – summers off, done at 3:30 in the afternoon. Teachers point out they don't even get a bathroom break or have access to a telephone.

So what can be done in the school house as we wait for the nation to commit to education as a national priority? We as educators can make sure that every child is understood and helped to learn critical skills at every grade level because "every mind matters".

- Provide information to parents of pre-schoolers about what is really expected of their children in kindergarten. Encourage them to read to their children daily and to help their children learn to listen and follow simple directions.

- Reach out to daycare providers and preschools to share information and resources.

- If a first grader is still struggling with reading after receiving extra reading support such as the popular Reading Recovery program, talk with the parent and bring the child to the attention of the school team that assesses student learning problems.

- Know the strongest predictors of reading achievement: alphabetic knowledge, phonological awareness (being able to analyze parts of words), rapid automatic naming of letters, digits, objects, and colors, writing one's name, phonological memory (being able to remember spoken information).

- Understand other predictors of literacy: knowledge of print, reading readiness, oral language development, skill in visual processing.

- Ask that various screening evaluations be used to determine why the child continues to struggle. Your school counselor should be able to provide them.

- Keep work samples to show your concerns. Share them with other teachers, the counselor, assistant principal, or school psychologist for input on possible learning problems.

- Before recommending a child for retention, understand that research does not support that retention results in any long-term achievement gain[44]. Numerous studies over many years that compared retained and those not-retained students with similar levels of academic achievement show negative effects on retained students in both achievement and adjustment. Controlling for level of underachievement, retained students are much more likely to eventually drop out of school than those who are not retained. There is also no long-term academic advantage to early versus late retention.

- Note too that social promotion is not the answer[45]. Those who are socially promoted without the skills necessary for success are also primed to drop out or graduate without the skills they need for today's job market.

What's A Teacher To Do?

- So what's a teacher to do? Abandon the debate between retention and social promotion urges Shane Jimerson in his recommendations regarding retention in the 21st Century[46]. He encourages early intervention, careful progress monitoring, and research-based alternative strategies to ensure that children are learning key skills at each grade level.

- Instead of settling for retention or social promotion, pull together a team of staff people to develop strategies to review the records of these children to determine how to help them succeed. What to look for? Attendance problems. English Language Status. Health and developmental problems. Previous grades and reading level. Writing samples. Learning style. Past successes and future options.

- If you feel you must retain – or your district's policies insist on it - how will the child's program differ the second year in the grade? If you are putting the child back through the same program that didn't work the first time, what are you hoping to accomplish?

- Embrace approaches like Response To Instruction (RTI) and Professional Learning Communities (PLCs) that professionalize teaching and individualize the teaching/learning process. Establish universal screening and progress monitor for struggling students to determine the type and level of support they need.

- Begin third grade by having each child read aloud to you. Check for fluency and phonics skills. Children who still haven't *learned to read* can not *read to learn*. They can not solve math problems if they can't read them.

- Bring struggling third graders to the attention of their parents and school-based assistance team to develop strategies or consider further evaluation. Consider a "pyramid of interventions" approach in which more intensive resources are provided to struggling students so no child's problems are overlooked.

- Consider administering learning style inventories. There is a wealth of data to support that addressing students' learning style needs results in improved achievement.

- At the beginning of fifth grade have each child read aloud to you – again. My own child's fifth grade teacher did this each year and each year she found children who were still struggling with basic reading skills. How could students get this far with a reading problem? How will they keep up in middle school if they aren't fluent readers? How much longer in their school career can they wait to get the reading help they need?

Make Time For Science And History

- Make time for science and history – even if they aren't on the high stakes test list. We are short-changing our children's futures when we don't teach the content of these core subjects. Have children read historical fiction. Use history and science content to teach reading skills. Use geography and science vocabulary for spelling tests.

- Insist that children remember facts, memorize, and answer in complete sentences. How can reading comprehension develop when our children have no learned information from which to draw[47]? How can our children function in a global society if they can't use a globe?

- Recognize the trouble makers, class clowns, those who say everything is boring, the quiet, brooding ones, for what they are – children crying out for help. No child wants to fail. They don't want to be embarrassed. They would rather be viewed as "bad" than "stupid".

- Monitor children's progress quarterly in core subjects using common assessments throughout the grade level or in all sections of the subject. Who is not mastering the third grade math curriculum? Who is not on target to achieve the needed level in reading? Who is falling behind in algebra? What will be done about it?

Be On Top Of Your Game

- Be as prepared as one of my favorite middle school teachers always is for parent meetings. "I have my data!" she says as she defends her educational plans for a struggling student. A boomer like myself who could have retired, at least mentally, she charts every student's progress – every day – and uses the data to plan each child's instruction. No kicking back, no lesson plans from the 70s, she's on top of her game every day.

- Consider grouping upper elementary students by academic need rather than age or grade level – another way to end the debate over retention. Those who are still struggling with decoding meet together each day for intensive reading instruction. Flex group for math so that those who still can't add fractions accurately get the specific help they need before moving on. Yes, it would be hard to do. So?

- Understand the process of language acquisition for your non-English speakers. Older students may have had limited time in school prior to enrollment in the U.S. Young students may have not learned the rules of their own language – a critical component for second language acquisition.

- Many second language learners go through a "silent period" of days or weeks before they will speak in class. If they are also not speaking with others who share their native language, ask your language specialist for advice.

- Remember that some older students have had very limited educational experiences. Having someone translate "compare/contrast" into Spanish is not helpful if the child has never compared or contrasted two reading passages before.

- Don't mistake lack of language for lack of ability. Compare a struggling non-English speaker with peers who have a similar time in U.S. schools. If the child remains far behind such peers, seek assistance.

- Don't confuse conversational English with knowledge of English that is adequate for learning. Non-English speakers may take from 5-7 years to acquire academic English. Just because you could find a bathroom in Germany doesn't mean you could learn chemistry in a German classroom.

Admit Your Prejudices

- Admit your prejudices. We all have them. We all silently make judgments about "those" children. Smart aleck gifted kids who shouldn't be getting any extra help since they can learn anyway. The ghetto kid with the baggy pants who won't answer or even look at you when you speak to him. The trenchcoat-type that you let sit in the back of the room and doodle disturbing pictures. The pimple-faced, heavy-set girl with glasses who rarely turns anything in. Those kids who shouldn't be in America and can't even speak English. Each of those children is just that – a child. Each was once a little baby at the mercy of the environment and now depends on us to see beyond the rough or rowdy exterior to the little boy or girl who still lives inside.

- Look through the cumulative folder – that vast repository of shot records, grades, attendance, and discipline – to understand your students' school history. I was shocked recently during a 9th grade reading screening. A vivacious young lady single-handedly coordinated the entire screening process for me without being asked but ended up scoring far below grade level in word recognition. I was further shocked to look in her cum folder to see she had already been retained twice and was a 16-year-old freshman. How could such a bright-eyed, engaging, take-charge personality have already flunked two grades and be getting ready to fail 9th - and how could her teacher not know any of that? Sounds like a bright young lady with some type of undiscovered learning disability to me.

- Ask yourself if your school's discipline policy is "fair" and "consistent" at the expense of being "effective" or logical. Granted the consequences of rule infractions should become greater the more the student misbehaves. Still I wonder how suspending a child from school for skipping school makes any logical sense!

- If a student has been suspended several times for skipping school, is the suspension working to change the behavior? Apparently not! Might a referral to a school team who reviews the child's grades, attendance, and discipline record and creates an action plan be more effective?

- Consider a "therapeutic" in-school suspension (ISS) program. After several ISS referrals, provide a therapeutic setting – not just a heavy hand and loud voice to make the student sit down and be quiet – to assess the student's real needs. Can he read? Is she struggling with a home problem? What's the point of ten days in ISS? What's the cost of intervention versus failure?

Middle School Is Critical

- Create an intervention team in every middle school to identify and help every at-risk student. Who is at risk? Students with failing grades or dropping grades, poor attendance, and discipline referrals. While that would seem obvious, it is too easy for children to fall through the cracks of the middle school hallways.

- Every child who comes to middle school having failed a grade should be on a support list. It is very unlikely that these students are yet on grade level and the demands of middle school and increasing peer pressure are a lethal mix for school failure.

- Children at-risk for school failure also make fertile ground for recruitment into risky behavior and illegal activity. Don't ignore the signs of gang recruitment. Once they're in, you may have lost them.

- Insist that your middle school students perform all the basic math operations and apply them to solve problems – without a calculator!

- Do not retain a child for a second time in middle school! Remember the research! Most students with two retentions will eventually become a dropout statistic. Instead evaluate the child's learning needs and set up an action plan to get him or her back on the right track. The alternative is the creation of a drop-out in waiting.

- Make sure your counselors have time to counsel! What a concept. Most counselors are so busy with paperwork and testing they have little time to get to know and help students in crisis.

High School, Your Last Chance…

- If you are a 9th grade English teacher – you guessed it – listen to every one of your students read. This can be done in a way that does not single anyone out or embarrass anyone, but it will provide critical information about the student's potential for success in high school. I recently completed reading screenings on four 9th grade classes and found at least 25% of the students were reading well below high school level. As one boy squinted at the words I placed in front of him I asked about glasses. "I've needed glasses for years," he said angrily from the back of the classroom, "I can't even see the board." A call to grandma confirmed the need and the lack of money. Several weeks later with the help of the school social worker his vision is 20/20 – with new glasses.

- When you identify students still struggling to read, talk as an English department about what can be done. You hold a key piece of information that every one of the child's teachers for the next four (or five, or six) years needs.

- Raise expectations! As I sit in assistance teams to discuss student progress it is distressing to hear "he's passing my class" as an indication that no further help is needed. Is passing good enough, especially when students are spoon-fed content for the test to get that grade of 70? Ask yourself "why"? Instead of insisting that Tracy could pass if she just stopped talking, stayed awake in class, or did her homework, ask yourself why Tracy is making such choices. When and why did Tracy give up on her education?

- How will your non-English speaking students learn English? Do we really want to foster separate school cultures in which Latino students congregate and speak together in Spanish while white students ignore them, chattering away in English? Do we really think non-English speakers can master physical science and algebra with nothing more than a Spanish-English dictionary and some tutoring?

- Look at the failure list of any large high school and the names on the list will confirm the challenge facing non-English speaking students. Identify Spanish-speaking staff and give them the time and training to reach out to parents.

- Consider creating an advisory program at the middle and high school level. With every teacher also an advisor, each child has a "go-to" person who knows their grades, attendance, and "issues". As a parent at the high school told me in support of the school's advisory program, "I'm so glad somebody out here knows my whole child".

- Advisories can meet daily, weekly, or monthly, with or without a specially designed curriculum. Topics of discussion can include decision making, coping with peer pressure, and study strategies. Provide teacher training and support for this new role.

- If you use advisories and your school issues progress reports electronically, have those reports sent to each student's teacher advisor. Nothing like one person at school realizing that Johnny is failing every subject and working with him to turn things around! Nothing like the pride teachers as advisors feel when Johnny finally walks across the stage to get the diploma he might not have obtained without their help.

- Students actually begin dropping out in elementary school. We have usually retained them often enough that they can formally drop out in ninth grade at the state mandated age of 16. Thus ninth grade becomes a critical year as students routinely fail English, earth science, world history, and algebra. Would you stay in high school as a fifth-year freshman? What does your school do to identify and help such children?

- Amazingly a young lady I recently evaluated had stayed in school as she failed Algebra over and over again. She tearfully asked me "Why can't I pass my grade?" How had this child's math disability not been identified before? A conference with her counselor and a change in schedule made her a junior and gave her a boost of confidence that she would eventually graduate.

- As I watch students slowly walking to the local high school in the morning, cigarette in hand and ignoring the tardy bell, I see students who have given up. What about the students who, after getting dropped off at school, immediately pull out their cell phone and head off in the opposite direction? Could that be your son or daughter I see? Consider special ninth grade programs designed to get these students reconnected.

- Offer credit recovery at the high school so students can hope to graduate. Expand summer programs and after-school academies to build missing skills. Consider "fast-track" programs with fewer credit requirements to provide students a path to success.

- Expect more, not just good enough. We the parents are allowing our children to do less than they could. We as educators need to re-evaluate our grades in light of what these students will need to compete. Is that really an "A" paper? Does that 10-point physics essay really deserve 10 points?

Do The Right Thing, Every Day

- Greet students everyday by name, and with a smile – even when you don't feel like it. For many students school is the calmest, safest, warmest place to be in a sea of family and community stress. Help them see that their lives can be different.

- Sarcasm doesn't work. Criticize the behavior but never the child. As the saying goes, "they don't care what you know until they know you care". Students are experts at body language and facial expressions – you can't fool them. They respond to teachers who offer respect and high expectations and will react in kind to those they think look down on them.

- Don't retire mentally before you retire physically. When you stop looking forward to setting up your room in August, it's time to hang up your apple.

- Help students see that there are different rules for different roles in life. Fighting and harsh language may be accepted at home or on the street, but not at school. I recently worked with a fifth-grade, African American boy who was being served in a classroom for emotionally disabled students. What I found was a bright young man who wanted to learn to read and spell. As we established some rapport, and he began to believe I wouldn't just write him off, he told me of his academic problems and his frustration that he wasn't learning anything. He understood that fighting was not the way to go even though in his previous environment it was necessary for survival. He agreed to try to restrain his temper if he could move to a class that would help him learn to read. The school team was proud of his accomplishments and he's currently visiting regular classrooms as a learning disabled – but not emotionally disabled – student.

- Teach parents how to be your partner in educating their children. If you've ever had to write an algebra equation to help your child with 5th grade math you'll know what I mean! Consider math night for parents to present grade level curriculum, translate "old math" to "new math", offer math games, and recommend good computer software.

- Have a read-in with parents and children reading to each other. Create displays of grade level books. Have a "make and take" for flash cards of sight words and vocabulary games.

- What about difficult parents? Those who think you are picking on their child, just not teaching correctly, or who won't admit there's a problem. Smile and document! Keep your good humor and don't blame the child. Document your concerns and ask others to observe and comment. Keep teaching and trying until the last day of school knowing you've done everything you could.

- Renew yourself to renew your professionalism. Don't give in to frustration. Don't let the paperwork bury you. Don't let the naysayers in the teacher's lounge get you down. Remember the teacher who changed your life – the one who led you to teaching. You are that teacher for someone, perhaps many someones, so don't give up!

Upon the subject of education, not presuming to dictate any plan or system respecting it, I can only say that I view it as the most important subject which we as a people may be engaged in...

President Abraham Lincoln

12

In the State House

Think Tanks, Policy Groups, and A Task Force –
Oh My!

The deep concerns about America's competitiveness in the new global economy have spurred action across the fifty states. Throughout the nation governors, legislatures, advocacy groups, and think tanks have set up task forces, created new policies, and funded programs. These initiatives will have the most direct impact on you and your children as new programs and standards are considered.

Whether your children attend a public school, magnet school, private school, charter school, or home school, you have a vital interest in understanding your state's approach to education reform. It is critical for you to review the education promises of the governor, your state legislature, and local governments with this question in mind: what do you propose to do to make sure education remains a top priority? What are the proposals you advocate and will implement to address the four principles of equal opportunity, individual freedom, innovation, and promoting a rich and vibrant culture in this, my home state?

In 2007 the National Governor's Association laid out a plan called Innovation America[48]. The initiative was led by then Arizona governor Janet Napolitano who noted, "United States economic growth in the 21st century will be driven by our nation's ability to innovate". The proposal provides both short- and long-range strategies including working with business on policies to promote innovation, strengthening the "STEM" subjects of science, technology, engineering, and math in K-12 education, and better methods for using new knowledge to strengthen local economies. The report notes that governors' roles are critical since states are the major investor in our human resources through K-12 education, community colleges, and universities in their states. Here too we can recognize the governor's role not only in spurring innovation for job creation, but insuring equal opportunities, protecting the freedom of its citizens, and in enhancing the cultural opportunities of all citizens that so contribute to our sense of well-being.

At the K-12 level, we continue to slice and dice education statistics as states consider what steps to take to improve education. According to Innovate America, one-third of the nation's eighth graders have less than a basic level of math skills and two-fifths have less than a basic level of science skills. While there have been across-the-board improvements in the skills of the nation's children, our relative standing compared to the rest of the world has fallen.

Even at the collegiate level the number of students gaining degrees in math, physical sciences, or engineering is far below what it was 20 years ago. In fact 55% of new engineering Ph.D.s and 38% of physical science Ph.D.s go to foreign students on temporary visas. The opportunities being created for them through the global economy suggests they will take the training they attained in America's finest universities and go home to compete with us in the global economy.

The State Of The States On Education

The National Governor's Association has also called for "benchmarking" or aligning our students' education with those of our competitors around the world. Indeed the states and the governors who lead them have a tremendous impact on your child's future. State education departments, in conjunction with governors' education policy advisors, decide how much math and science your child should take in high school. Twenty years ago some students may have only been required to take one math and one science class to graduate, now most states require at least three, often four.

States also determine teacher qualifications and allow – or forbid – teachers in classrooms who are not trained in the subject they are teaching. The qualifications of your child's algebra teacher depend in large part on who gets elected in your state – and who gets elected is up to you!

States also initiate – or choose not to initiate - specific programs aimed at reducing the dropout rate, encouraging girls and minority students to take more math and science, providing technical college classes in high school, and a myriad of other programs that could offer just the help your child needs to succeed. Do you know where your elected state officials stand on these critical issues? State lawmakers in Colorado previously rejected a proposal to require four years of math and three years of science of all high school students. Whether they were concerned about having to cut art classes or finding qualified teachers, do you as a Colorado parent agree with your legislators' decision? Did you even know about it?

Some states are offering specialized training for teachers to improve their ability to teach the skills of the 21st century to a diverse student body. As anyone who has ever suffered through a class with a poor teacher or professor knows, a great teacher presents challenging content in an accessible way. Is your governor talking about teacher training? Does your school district provide quality teacher training? If not, why not?

Other states have harnessed the resources of American business and federal grants to provide technical training and help to restructure high schools. Governors are key to this effort. I worked with one such federal grant to create "career-focused learning communities" and "academic advisories" at a local high school. The effort was to provide rigorous courses, help students see the relevance of high school to their future career, and to make sure every student had a campus mentor who could shepherd them through. Parents were critical to making things happen in the program through becoming career mentors, lobbying the school board for new programs and equipment, and for insisting their children participate. Students visited college campuses for engineering day, studied the components of leadership, interned at local hospitals, ran the school's computer network and website, and produced the daily school video news.

States can take on other initiatives as well. From math and science camps, scholarships, summer academies, and improving technology availability, your state is a vital resource to you in making sure your children have a place in the new economy.

Your state also decides what requirements it will place on homeschooling. Must you have a degree? Must you follow a specific curriculum? What are the attendance and testing requirement to be? States also regulate charter schools. How many are allowed across the state? How are they evaluated? What programs and services will be available for your special needs child? What services are available for severely emotionally disturbed or delinquent children?

Your school district has an even more direct impact on your child's future. Will your child be bussed or redistricted? How many psychologists are available to assess your child's learning problems and get them help if they need it? Does your district recruit the best teachers – and keep them? Does your high school offer advanced courses, enough foreign language classes, enough computers? Do you feel your child has the same opportunities as any other child in your district? Do you feel free to express your opinions? Are your family's cultural traditions celebrated, or at least recognized? Have you ever been to a school board meeting where such decisions are made? Who are your school board members anyway? Do they know what you think? Do they care?

A recent public opinion poll sponsored by the Governor's Association found that a clear majority of Americans do believe that states play a key role in encouraging innovation through education. Is your state making it happen? The following summaries show what was supposed to happen according to the National Governor's Association[49]. Many states became part of the American Diploma Project (ADP) to align high school, college, and workplace readiness standards – a critical component of America's Renaissance. In the last year significant federal funding has flowed to certain states, districts, and public/private ventures for innovation in education. Many states are actively working to make a difference. Is yours? The money carrot came with a policy stick to encourage teacher evaluation, charter school expansion, innovation in technology, acceptance of national standards, and attention to both early education and high school graduation rates. So...

What Has Your State Been Up To?

Alabama: Alabama was the first state to require four units of math and science for graduation. Beginning in 2009 entering 9th graders must complete an "advanced" diploma which will provide its students with one of the most demanding math programs in the country. The state has also invested in teacher training and on-line technology to bolster math and science. It rated 75.3 on the Quality Counts report and shows an abysmal 62.5% graduation rate.

California: Apparently California will have to wait for a needs-based grant from the feds – they didn't win the Race to the Top. Broke and showing a 62.7% graduation rate, the state has school districts where the majority of students don't even speak English. Still LA Unified is receiving an Innovation grant for a "portfolio of high performance schools" geared to community needs and offering open competition to the private sector to turn around failing schools.

Colorado: (ADP state), created a P-20 (pre-school through college) Education Coordinating Council. Its goal was to cut the state's dropout rate in half and double the college degree and technical certificates in the state. Colorado also promotes a "response to intervention" model, even among its high schools, that monitors student's progress closely and ties increasingly intense interventions to promote student success. They recently overhauled teacher tenure and developed a smart statewide framework for gauging teacher performance but somehow didn't make the cut for Race to the Top causing the state's lieutenant governor to question the objectivity of the process. Innovation grant funding will help their St. Vrain school district create a unique literacy, science, and math program for Hispanic and other ELL students however.

Connecticut: (ADP state) In 2005 the governor facilitated a statewide initiative of business and education leaders to improve K-12 student achievement and set up the Early Childhood Research and Policy Council. The state's Business and Industry Services Network makes sure that what children are learning is what business is looking for. It scores 72.5 on Qualify Counts with a 73.2% graduation rate.

Delaware: (ADP state) Delaware's governor focused on a Career and Technology Education group working to align K-12 education, industry, and college. Graduate teaching fellows work with high school science teachers to improve instruction. The state won big bucks in Round 1 of the RttT competition for federal dollars. Hope it improves their graduation rate of 65% and closes the 20.7% of its high schools deemed "dropout factories".

District of Columbia: Under the controversial reign of reformer Michelle Rhee, DC won a Race to the Top grant of up to $75 million. With a Quality Counts grade of 68.3 and graduation rate of 59.5, there is nowhere to go but up in the nation's capital. Will election year politics derail reforms?

Florida: Enhancing STEM programs was an initiative of Florida's governor, driving new world-class math and science standards in the state with quality teacher training programs. Florida, too, won big in RttT yet has one of lowest graduation rates in the nation at 62.1%. The Miami Dade district won Innovation dollars for early childhood programs.

Georgia: (ADP state) The governor worked to bring internet access to poor and rural counties throughout the state. This provides educational opportunities and a better quality of life to its residents. Georgia has also strengthened its K-12 curriculum and won a $400 million Race to the Top grant yet continues to struggle with a 57.8% graduation rate based on the 38.9% of its "dropout factory" high schools. Forsyth County's Innovation grant for a data system for personalized education shows promise.

Iowa and Kentucky (ADP state): Project Lead The Way, a four-year sequence of courses of introductory engineering topics, is supported by governors in these states. Project Lead The Way is now found in high schools around the country to encourage students to consider engineering as a field of study. Iowa can brag about its 80.2% graduation rate and Kentucky can hope for Jefferson County's new "trimester high schools" offering credit recovery, more electives, and advisories.

Massachusetts (ADP state): The governor proposed a requirement for three years of math and science in high school. He also champions ongoing teacher training. The state has also chosen to measure itself against other nations, one of only two states to do, and came closer to international top scores than other states. They received the top score in round 2 of Race to the Top and rated one of the few true "Bs" on Quality Counts.

Michigan (ADP state): Michigan's governor has made the state one of twelve to align high school graduation requirements with college and workplace needs. Michigan also became the first state in the nation to require an online learning class for graduation. A partnership with Microsoft provides a free online career planning course to Michigan students as well.

Minnesota (ADP state): The governor created the Achieve Scholarship that provides college funds to students who complete rigorous high school courses. He also required accountability reports from the state's higher education institutions. Minnesota is the second state, along with Massachusetts, to judge itself against other nations. It too is scoring at or near the level of the highest performing countries on math and science.

New York: A surprise to some, New York garnered a Race to the Top grant although it scores low on many national rankings. It tied with Florida as the largest RttT grants at up to $700 million. It's "School of 1" program which uses technology to adapt to each student's skills, interests, and learning preferences garnered an Innovation grant as well. We should see big things from New York, right?

North Carolina (ADP state): The former governor launched a large high school innovation initiative with support from the Bill and Melinda Gates Foundation. The state has created nearly 90 "Learn and Earn" programs in which students take a five-year program of high school and college courses earning them an associate's degree and/or college credit. The new governor vows to make education a centerpiece of the state's renaissance as well. Race to the Top funding should help, and will emphasize STEM programs and innovative high schools to help the state's 57.8% graduation rate and its high school "dropout factories".

Ohio: The Buckeye state was another surprise winner of Race to the Top funds. Conservative ed analyst Chester Finn and Democratic education pundit Andy Rotherham both questioned the winners and losers in the RttT race. Both on charter policy and teacher quality, Ohio brings in disappointing grades yet may have up to 400 million RttT $ to bring those grades up.

Rhode Island (ADP state): Students here may soon all be required to take physics, chemistry, and biology, in some cases via the internet! The governor has a STEM action plan that includes higher standards that integrate with the needs of business. A Race to the Top winner, they'll be getting up to $75 million.

Washington (ADP state): The governor set up four-year scholarships for students who score at level 4 of the 10th grade math or science assessments. The state has also created a Department of Early Learning to look at ways to infuse math and science topics early in a child's educational life. With an Innovation grant, high schools in Bellevue are shifting to a problem-based curriculum to encourage under achievers.

So, who sits in the "state house" is critical to your child's future. Scholarships, laptops, summer camps, better trained teachers, college coursework in high school, accessing federal support, all may be available to your child when your state and local officials "get it" about the need to make education "Priority One".

Even if you have chosen a different "place" for your child's education, the state house is key. States regulate charter schools and homeschooling as well. No matter what path you have chosen for your child's education, who sits in the state house matters.

...The problem I have is in the implication that if it's hard, don't do it...

Ken Kay, P21 Group for 21st Century Education Standards

13

In the White House

The election of 2008 is not only "history", it was *historic* with Barack Obama's election to the presidency. Still, as the economic crisis rose to the top of the nation's agenda, less was said about specific and practical proposals for making **Education Priority One in America – even though education is critical to America's long-term prosperity – and maybe our survival as a free and democratic nation.** While rebuilding crumbling classrooms and fixing leaking roofs will help, it's not enough.

While the federal government typically provides only about 10% of school funding, it provides a much higher percentage of the research that leads states to action. The federal government funds the specialized reading programs that benefit struggling first graders. It funds the research into math education that will influence everything from teacher training to math textbook selection. The federal government funds some of the programs for children with special needs and most recently prompted an unprecedented level of testing and accountability.

No matter who sits in the White House, the failings of American education must be addressed. Although my upbringing in the American heartland laid the foundation for a conservative, hands-off my life and business approach to the federal government, I've come to believe that it's time to lay ideology and party affiliations aside. Although it's true the

Constitution does not grant powers to the federal government regarding *managing* education, leaving those powers to state and local government, the Constitution does require that the federal government do many other things that are impossible without an educated populace:

- form a more perfect Union
- establish justice
- insure domestic tranquility
- provide for the common defense
- promote the general welfare
- secure the blessings of liberty

Thus the Constitution lays the perfect foundation for the principles of equal opportunity, freedom, innovation, and rich culture. The Bill of Rights goes on to guarantee these values. "Leave it up to state and local government" just doesn't cut it any more. There is too much at stake. That's where the "right" is wrong! We've been increasing education spending for years and we're still leaving millions of children behind. That's where the "left" is wrong. It's time for middle Americans to do what's needed in their own house then mobilize the federal government to do what's needed in the White House. With the global competition, economic crises, terror threats, and environmental challenges America faces in the 21st century, it is clear that our children are not yet receiving the education they need to uphold the goals laid out by the Founding Fathers so many years ago.

How can education help form...

A more perfect union...

There is already some concern that our "union" is not very "united", and certainly not yet perfect. The red state/blue state divisions, the new "tea" party activists, our knack for labeling, name calling, and dividing everyone, the tension over immigration, all detract from our unity as a country. It is only through education that young people will learn the facts about these issues and begin to resist "education by sound bite".

Establish justice...

We've all seen the marches and slogans "no justice no peace". Fortunately we Americans have the right to insist that our grievances be heard by the government. When we see injustice we have the freedom to speak out about it. Still, without an educated populace, we are easily swayed by this or that advocacy group, sound-bite explanations of complex issues, or the latest "ruling" on court TV.

Do our children understand their rights under the constitution? Do they understand that billions of people around the world have no such rights? When my own children join the Right To Life March in Washington DC each year I make sure they know they are exercising several of their rights as Americans. The federal government must continually be held accountable for preserving these rights, some of which are under attack as common citizens lose privacy so the bad guys don't get away.

Insure domestic tranquility...

How tranquil can America be with potentially millions of young people who are unemployable in the new economy? How tranquil will your family be if someone loses a job and can't secure a new one at the same salary? How will your son or daughter pay off the rising cost of a college education if he or she can't get a job? If the differences between the haves and the have nots continue to increase, might not the strikes and riots of the past come back to haunt America?

Provide for the common defense...

Here's an area of federal responsibility on which people of all political parties tend to agree. America has always championed freedom around the world and protected its homeland against threat. The question today seems to be what constitutes the "common defense"? Does it include removing tyrants from government in other countries? Does it include maintaining troops in countries around the world? Can America responsibly remove itself from world conflicts in the hope that faraway battles will not reach our shores? Must we be afraid of home-grown terrorists, of our neighbor next door?

Here again education is key. Not only can most students probably not find Iraq, Darfur, or Korea on a map, they have no idea of the enormous issues at stake. As a boomer I remember well the nightly body counts during Viet Nam, and holding my breath as the lottery numbers came up to see who from my hometown would be the first to go. I remember singing the protest songs and still get teary-eyed thinking about the guys who didn't make it back, the fear they must have felt in that faraway jungle, and the homeless vets who still fight those battles in their minds to this day. Not only must our children understand the history of America's wars, they must have the education and training necessary to fight America's next battles as well.

Promote the general welfare...

...that's a pretty broad mandate from our Founding Fathers. The dictionary defines welfare as *health, happiness, and prosperity* – the very aspects of the economy many Americans are so frightened about. These are also areas for grand election year promises, programs, and policies – many of which are often forgotten when the winner takes office and reality bites. If it's not defense it's the "economy, stupid". Whatever the election buzzword, education is still the key. Not only does it take an educated voter to sort through proposals on healthcare and taxes, it takes an educated American to take care of his own health, maintain a problem-solving, can-do attitude, and ensure his own prosperity with enough resources left to help those in need. The statistics on the costs of the nation's dropout rate also remind us that adding those billions of lost dollars back into the economy would go pretty far toward "promoting the general welfare".

Secure the blessings of liberty for ourselves and our posterity...

The blessings of liberty - and what blessings they are. Our children are growing up in the wealthiest, freest, nation in the world. No country offers more opportunity to more of its citizens. Even the new economic powerhouses of China and India leave millions of their people behind. China still silences dissent and India's caste system still wields its subtle power. While clearly there are still invisible – and some very visible – barriers to the blessings of liberty for all Americans, again education is the great equalizer. If you have the education and training for college admission, technical training, or that first job or promotion, your gender, race, age, or handicap can not be used to stop you. If you aren't qualified, if you can't read the manual or "do the math", the barriers you face will be insurmountable.

So how do we create an American Renaissance? How do we make Education Priority One in the White House? By having Americans of all political parties agree that education *is* a national responsibility and a national priority and demanding that our national leaders in the White House and Congress respond accordingly. Whether you're staying with the two-party system, joining some Independent movement, or considering dumping some tea, come together to promote education. Insist that our elected officials positively promote equal opportunities, freedom, innovation, and our rich culture. Hold the president and elected officials accountable for their promises. Remember that the president has the "bully pulpit" to rally the Congress, business leaders, governors - and you - to invest in the education of our children.

The following are some questions to ask of the leader of the free world as he continues his term:

- It is clear that some children arrive in school at a disadvantage. How do you think that should be addressed? If you are proposing a pre-school program, what are you expecting three- and four-year-olds to be able to do? Will the program be mandated? Free? Subsidized? Is it the government's job or the parent's job to provide this education?

- What is your position on school choice? What policies will you pursue to ensure that quality education is available to every child in every place in America?

- What educational research is most critical to complete? What is it that we don't know yet that we need to know to do a better job of educating our children?

- What is your position on why, how, and when our immigrant children learn English? What programs will you promote for English Language Learners? How will you help states deal with the costs of educating these students?

- What specifically will you ask the business community to do to improve public education?

- What are the additional educational initiatives you will undertake in office? A longer school year? Changes to teacher accountability? How will education have improved by the end of your first term?

- What is the role of the federal government in ensuring that college is affordable?

- How do you propose to change the No Child Left Behind legislation? Will you maintain an end-of-year high stakes test or move to ongoing assessment of individual student progress? What about science and history? How will we know our children are learning those key subjects? How can the progress of my disabled child be fairly evaluated?

- Can we count on you to keep sounding the call for personal responsibility? You said it takes each one of us to make a difference, that you can't do it alone. Will you keep prodding us, reminding us, and calling on each of us take charge of our own families and our children's education?

Ask these questions – and more – and demand explicit answers. If we all ask the questions, and ask them often, the leadership in America will use the support of millions of concerned Americans to begin to make things happen. Keep asking the questions and demanding action! Call, write, e-mail, and fax the people you elected to make sure they are following through with the promises they made. Visit www.whitehouse.gov often to stay informed and offer your thoughts. Find out how your representatives voted on key education issues.

Don't let promising pieces of education legislation get lost in a sea of corporate bailouts and pork barrel projects. It could mean a lost opportunity for your own child's future. You can't take your private jet to Washington to ask for money like the auto execs did but you can call and write and email!

Use state associations such as the PTA, homeschool association, religious organization, business roundtables, etc. as action committees to speak with one voice for education. If your state won recent federal grants, hold them accountable for the results. Don't let up until we can all be confident that our children won't be left struggling to find jobs, afford a home, and survive in the coming decades.

14

Education –
It's Mind Altering!

America is at a critical juncture. The America in which you and I grew up has changed. The whole world has changed as the balance of power starts to swing from West to East. Billions of people are competing for jobs, food, clean water and air, and a safe place to live. Information is instantly available and decisions that used to be pondered for weeks are now made in a split second. Cultures that have collided for centuries now face each other daily with competing world views. Minorities have become majorities in parts of America yet often without the education they need to succeed and contribute to a stable and prosperous society.

Shall we simply accept the inevitable? At least there's a new episode of our favorite crime show or comedy on tonight right? We've got our 50-inch screen for the big game, right? Aren't we really awfully busy – too busy to worry about things we can't control anyway? Nothing we can do about it, right? Wrong, wrong, wrong!

If you have children you can do something right this minute! Sit down and talk to them. Read to them. Pull out the backpack and empty it right on the kitchen table and see what's going on! Talk to other parents. Visit the school. Re-evaluate the educational choices – yes you have choices – you've made for your child. Are your children receiving the best education possible? If not, then, *do something*. You're the only parents they've got. Rally other parents. Have a community meeting. Attend a school board meeting. See what charter schools are available in your area. Visit a local home-school group and talk to the parents about how they manage. Research private schools to see which ones offer scholarships. Take charge of your child's education. Make it priority one in your house and insist that it become priority one in everybody's house!

And A Word About... Other People's Children.

Just as in the 20s and 30s, those with means tried to help "other people's children" survive the Great Depression, so today we can not afford to lose a single child to ignorance. **Every mind matters!**

Former U. S. Secretary of State Colin Powell and his wife Alma started the "America's Promise Alliance"[50] to draw attention to the alarming dropout rate in America. They join other foundations, civil rights groups, non-profits, churches, and entertainment icons like Oprah Winfrey and Bill Cosby to motivate parents and take action to reach potential dropouts and keep them in school. They are all willing to commit time and resources to other people's children because those children are America's hope and inspiration.

What Can You Do?

What can you do for other people's children? America's Promise and other groups like it offer many ideas:

- Offer to mentor a student in your local school. Roughly 20% of our young people don't have a caring adult in their lives yet such an adult is a key factor in student resilience in the face of family and community risk factors.

- Invite teachers to your place of business to see the skills needed by your workers.

- Invite a teacher to dinner! Remember when it was common for teachers to visit their children's homes? Really listen to teachers. Ask them about their day, their students, their resources, and their frustrations. Get a clue about what is really going on in America's classrooms.

- Provide training and policy manuals from your business as a way of showing students the reading requirements of today's jobs.

- Allow your children's friends to share your family's activities. You can't be their parent but you can show them how families can function in support of each other.

- Provide a safe place for children who have no other options. Check on the latch-key child next door. Be the "safe place" in your neighborhood for children who are alone or afraid.

- Volunteer for an organization like Stand Up For Kids (www.standupforkids.com). Over one million children are living on the streets of America. Others are "skating" from house to house with no stable family support. Aren't these just delinquents who leave home rather than follow the rules? No, they are America's children, many of whom have often been victimized by society – or their own parents – and run away to escape.

- If you are in a medical profession share your expertise with your local schools. Find time to provide a classroom activity or lend some real-world experience to a high school biology class to encourage students to consider careers in medicine.

- Offer to tutor students in your area of expertise. Many parents can not afford the tutoring their child needs and your help could make all the difference.

- Involve the teens in your church or civic group in activities that contribute to the common good.

- Offer to participate in career fairs in your local high school. No career fair? You organize it!

- Be a grandma or grandpa to a child at your local school. Many children today don't have the benefit of a multi-generational family of caregivers. Offer to read to a child or lead an after school activity for children who need the loving attention of an older adult.

- Even if you don't have a child in public school, your future depends on those children. Stay up to date on school issues, insist that education remain a priority in your state, and don't automatically vote "NO" on school bonds!

- Want to stay fit and get some exercise? Coach a children's sports team, teaching them the value of hard work and sportsmanship.

The list is endless. Everyone can – and must – do something for America's children. From your house, to the school house, to the state house and the White House, the future of America and our place in the world of the 21st Century depends on you.

**What does the future hold for America,
its demise or its renaissance?**

That will be determined by you and me.

... The Future of America Is Education so
Education must be the Business of our Future...

Resources

Academic Support

Math and reading help for kids.
http://math-and-reading-help-for-kids.org/index.html

Parent information on reading from the federal government
http://www.ed.gov/parents/read/resources/edpicks.jhtml

"About" teaching English to non-English speakers
http://esl.about.com/od/englishreadingskills/English_Reading_Co
mprehension_Skills_for_ESL_EFL_Learners.htm

Interactive reading activities for students and teachers
http://www.globalclassroom.org/2005/inservice/reading.html

Interactive reading help for kindergarten and first grade
http://www.starfall.com/

A wealth of links for academic support and more.
http://www.internet4classrooms.com/parents.htm

National Institute for Literacy Parent Resource
http://www.nifl.gov/partnershipforreading/publications/reading_
first2.html

Helping your child learn to read
http://www.kidsource.com/kidsource/content/learread.html

Resources for reading help for your child
http://www.readingrockets.org/families/tutoring

Math instruction and homework help
www.math.com

Get help solving a tough math problem
http://www.webmath.com/

Help with high school math topics
http://www.freemathhelp.com/

Math help for middle and high school
http://education.yahoo.com/homework_help/math_help/

Get help with problems in your math textbook
http://hotmath.com/

Math help for grades 4-8
http://www.mathleague.com/help/help.htm

Find books at appropriate reading levels for your child
http://www.lexile.com

Learn about math skill sequences and how to help your child in math
http://www.quantile.com

Special Learning Needs

K-12 help for parents
http://www.schwablearning.org/

Wealth of information on learning disabilities
http://www.ldonline.org/

National center for learning disabilities
http://www.ncld.org/

A national directory of learning disability professionals
http://www.iser.com/

International dyslexia association
http://www.interdys.org/

Symptoms and treatment of dyslexia
http://www.medicinenet.com/dyslexia/article.htm

National Institute of Mental Health on ADHD
http://www.nimh.nih.gov/health/publications/adhd/complete-publication.shtml

Center for Disease Control and Prevention on ADHD
http://www.cdc.gov/ncbddd/adhd/

Parent support organization for ADHD
http://www.chadd.org/

Autism Society of America
http://www.autism-society.org/site/PageServer

Autism fact sheet
http://www.ninds.nih.gov/disorders/autism/detail_autism.htm

Autism speaks
http://www.autismspeaks.org/

Private schools

Private school reviews by state
http://www.privateschoolreview.com/

"About" private schools
http://privateschool.about.com/

Council for American Private Education
http://www.capenet.org/

National Association of Private Special Education
http://www.napsec.org/

The Montessori Approach and Schools
http://www.montessori.edu/

Homeschooling

Homeschool.com
http://www.homeschool.com/

Homeschool World
http://www.home-school.com/

Homeschool curriculum and ideas
http://homeschooling.gomilpitas.com/

Home School Legal Defense Association
http://www.hslda.org/Default.asp?bhcp=1

Homeschool Central
http://www.homeschoolcentral.com/

Essay on classical education
http://www.welltrainedmind.com/classed.html

Classical homeschooling
http://www.classical-homeschooling.org/

General

Example of state standard course of study
http://www.ncpublicschools.org/curriculum/

Teaching character
http://www.goodcharacter.com/

Intervention central
http://www.interventioncentral.org/

Community Support, Collaboration, Policy, Advocacy

Forum for investment in today's youth; all youth "credentialed by age 26"
http://www.forumforyouthinvestment.org

"Silent Epidemic"; the dropout problem
http://www.silentepidemic.org

Promote children and families in federal budget and policy
http://www.firstfrocus.net

Raising graduation standards, America Diploma Project
http://www.achieve.org

National School Psychology Association
http://www.nasp.org

Educational Data
http://www.kidscount.org

Educational News
http://www.edweek.org

Boys and Girls Clubs
http://www.bgca.org

Big Brother/Big Sister
http://www.bbbs.org

Middle and High Schools

Make a national investment in high school reform
http://www.all4ed.org

Pew Partnership for Civic Change
http://www.learningtofinish.org

The National High School Center
http://www.betterhighschools.org

Early College programs
http://www.earlycolleges.org

High School Redesign in North Carolina
http://www.newschoolsproject.org

Middle Schools Administrators Association
http://www.nmsa.org

High School Administrators Association
http://www.principals.org

Notes

1. Jennings, Peter, and Brewster, Todd; *The Century for Young People,* Scholastic, New York, 2000.

2. Steingart, Gabor, *The Depressed Superpower, Spiegel Online,* November, 2007.

3. Parris, Matthew, *Rusty Superpower In Need of Careful Driver,* TimesOnLine, January 3, 2009.

4. Ankarlo, Darrell, *What Went Wrong With America...And How To Fix It,* Cumberland House, Nashville, TN, 2004

5. Browne, Anthony, *Why China is the Real Master of the Universe,* London Daily Mail online, 4/13/2008.

6. Meredith, Robyn, *The Elephant and the Dragon: The Rise of India and China and What It Means for All of Us,* W. W. Norton & Company, New York, 2007

7. *Japan Fattens Textbooks To Reverse Sliding Rank,* Associated Press, 9/8/2010

8. Tooley, James, *The Beautiful Tree,* Cato Institute, 2009

9. Forsyth, Jim, *AT&T CEO Says Hard to Find Skilled U.S. Workers,* Yahoo News, 3/27/2008.

10. National Center on Education and the Economy, *Tough Choices Tough Times: The Report of the New Commission on the Skills of the American Workforce.* Jossey-Bass, San Francisco, 2007.

11. Strong American Schools report, *A Stagnant Nation: Why American Students Are Still At Risk,* 4/21/2008, www.edin08.com.

12. Organization for Economic Cooperation and Development, *OECD Economic Surveys: United States,* OECD Publications, 2007.

13. Strong American Schools Report: *Americans Deserve Leadership on Education,* www.edin08.com

14. Cutler, William W. III, *Parents and Schools,* The University of Chicago Press, Chicago, 2000.

15. Jonathan Kozol, http://www.cincinnati.com/samepage/author_kozol.html

16. Family Rights and Privacy Act, http://www.ed.gov/policy/gen/guid/fpco/ferpa/index.html.

17. Individuals With Disabilities Education Act, http://idea.ed.gov/.

18. James Comer, faculty page at Yale University, http://info.med.yale.edu/chldstdy/faculty/comer.html.

19. *Education Week* article on James Coleman, http://www.edweek.org/ew/articles/1995/04/05/28cole.h14.html.

20. *A Nation At Risk,* original article, 1983, http://www.ed.gov/pubs/NatAtRisk/index.html.

21. Overview of the law, *No Child Left Behind*, *http://www.ed.gov/nclb/overview/intro/edpicks.jhtml?src=ln*.

22. Magnet Schools of America, information on magnet schools, http://www.ed.gov/nclb/overview/intro/edpicks.jhtml?src=ln.

23. National Conference of State Legislatures, summary of high school reform, http://www.ncsl.org/programs/educ/HSReform.htm.

24. The Strong American Schools Project http://www.edin08.com.

25. Ed In 08, Making Education a Priority in the 2008 election, http://www.edin08.com.

26. *Education Week* publication: Diplomas Count, The Graduation Project 2010, Editorial Projects in Education, Vol 29, #43, 6/10/2010.

27. *Education Week* publication: Diplomas Count, The Graduation Project 2007, Editorial Project in Education, Vol 26, #40, 6/12/2007.

28. *Waiting For Superman,* Participant Media Guide, Public Affairs, 9/14/2010

29. Data 360 Report, *Trends In Public Education,* http://www.data360.org.

30. Siimonds, Robert L. and Hudson, Kathi, *Choosing Your Child's Future,* Citizens for Excellence In Education, Coata Mesa CA, 1998.

31. Council for American Private Education, CAPE, *Facts and Studies,* http://www.cpenet.org.

32. National Center For Educational Progress, http://nces.ed.gov/nationsreportcard/.

33. Bureau of Juvenile Justice Statistics, *Indicators of School Crime and Safety,* 2003, http://www.ojp.usdoj.gov/bjs/.

34. National Center for Education Statistics, *The Schools and Staffing Survey,* http://www.nces.ed.gov.

35. Christian School Education, *An Approach to Worldview Integration: A Key Teaching Tool for the Authentic Christian Teacher,* Vol 6, Issue 1.

36. *Fostering Innovations and Excellence,* Policy Blueprint, ed.gov

37. Home page of American Charter School Association, http://www.uscharterschools.org/pub/uscs_docs/index.htm .

38. Hoxby, Carolina M. *Straightforward Comparison Of Charter Schools And Regular Public Schools In The United States,* Harvard University And National Bureau Of Economic Research, September, 2004, http://www.mnohs.org/Harvardstudy.pdf.

39 National Center for Education Statistics, Study of Homeschooling in 2003, nces.ed.fov.pubs2006/homeschool/parentsreasons.asp.

40. Old Schoolhouse: The Magazine for Homeschool Families, *The Results Are In and Homeschooling Is Measuring Up,* Fall 2007, pp36 – 41.

41. The Education Reporter, *Wealthy, Suburban Schools Are Failing, Study Shows*, #263, December 2007.

42. Editorial Projects In Education, *Quality Counts2010*, www.edweek.org, 1/14/2010.

43. Keitel, Eirc & Cooper, Aaron, *I Just Want My Kids To Be Happy*, Late August Press, February, 2008.

44. Anderson, Gabrielle, Whipple, Angela, Jimerson, Shane, Center for Development and Learning, *Grade Retention: Achievement and Mental Health Outcomes*, www.cdl.org/resource-library/articles/grade_retention.php.

45. North Central Regional Educational Laboratory, *Beyond Social Promotion and Retention – Five Strategies to Help Students Succeed*, http://www.ncrel.org/sdrs/aras/issues/students/atrisk/at800.htm.

46. Jimerson, Shane R., School Psychology Review, *Meta-analysis of Grade Retention Research: Implications for Practice in the 21st Century*, Vol 30, #3, 2001, pp 420 – 437.

47. Hirsch, E. D. Jr., *Reading Incomprehension: Teaching Methods Ignore the Importance of Knowledge*, The Washington Post, 2/16/2008.

48. National Governor's Association Meeting Report, *Recent State Initiatives to Promote Innovation*, www.nga.org, 7/23/2007.

49. National Governor's Association, *Addendum: Recent State Initiatives to Promote Innovation,* http://www.nga.org.

50. The "America's Promise" homepage, http://www.americaspromise.org/APA.aspx.

www.ingramcontent.com/pod-product-compliance
Lightning Source LLC
Chambersburg PA
CBHW060624290526
45793CB00001B/134